The

Theology

of

BIOLOGY

War & Peace at the Molecular Level

EMUEL PAUL KIRBAS

Vine Press Publishing
Dallas, GA 30132

The Theology of Biology:
   War & Peace at the Molecular Level

Published by: VinePress Publishing, LLC
Dallas, GA 30132

We desire to hear from you. Please email us of your comments
to: comments@VinePressPublishing.com.

Publisher's Cataloging-in-Publication Data

Kirbas, Emuel Paul, 1930–.
   The theology of biology: war and peace at the molecular
level / Emuel Paul Kirbas – First Edition.

   xx, 372 p. : 22 ill. ; 23 cm.
   Includes bibliographical references and index.

   ISBN–13: 978-0-615-98509-1 (alk. paper).
   ISBN–10: 0-6159-8509-2

1. Bible and science. 2. Religion and science. I. Title.

BS 650 .K57 2014
215—dc23
2014904703 LCCN

Printed in the United States of America
20 19 18 17 16 15 14 / 1 2 3 4 5 6 7

DEDICATION

Lovingly dedicated to my wife, Caroline—my Sweetheart:
And to all my children; Joseph, Bonnie, Bryan, Paul, Kevin,
and Christopher.
To my grandchildren: Kristen, Corey, Parker, Bradley,
Brandon, Nicole, Taylor, Tassy, Nicholas and Sarah;
And to all my great-grandchildren—(whenever they will be).

Acknowledgments

I wish to acknowledge immediately the magnificent work authored, complied, and edited by Virgil C. Mayes, entitled *Leaves, Worms, Butterflies, and T.U.L.I.P.S.*, where the second chapter of this work finds its title name. In addition, appreciation is extended to many of the theologians of the past era, notably Arthur Pink, Charles Spurgeon, and Adam Clarke, to mention but a few, for their numerous groundbreaking books on the many sound doctrines of the Christian faith. Also to Dr. Henry M. Morris, of the Institute of Creation Research, for addressing the recent creation/evolution debate, and to Dr. S. Lewis Johnson, past pastor of Believer's Chapel, Dallas, Texas for his many audio sermons on the entire Bible that I have been blessed to have listened to over the past forty years.

Books by these authors should be on the bookshelves of all Bible-believing Christians of all persuasions. Other contributing authors are quoted extensively and the various contributions seem to have become part of the major theme throughout the volume of this book.

It seems to me as though the project of developing and writing this volume over the past twelve years was nothing short of a miracle, for I am an Electrical Engineer by trade and definitely not an author. I admit to everyone, therefore, that I was utterly incapable of writing a work such as this without the invaluable assistance of a host of dedicated professionals in the field of theology, biology, physics, neurology, and philosophy. So many good works from so many excellent authors that I have read during the past decades undoubtedly have found their way into this work without due recognition for the use of some of their helpful insights. Forgive me to have long ago forgotten your names, the titles of your books, and the extent of your contributions.

Both the Bible and the scientific community have been quoted extensively, from which various images have been derived. First, you will see something on the cover of the book that resembles the DNA molecule with its notable helix twist that extends from the top of the page to the bottom, as if coming out of heaven itself, with its two side rails and connecting rods. This also relates to a biblical passage in the book of Genesis where Jacob, grandson of Abraham and Sarah, is sleeping under a tree, dreaming about a staircase, or ladder, extending from heaven to earth, with the angels of God ascending and descending upon it; known as Jacob's Ladder, where Jesus uses the ladder metaphor saying: *"Hereafter ye shall see heaven open, and the angels of God ascending and descending upon the Son of man"* (John 1:51). This is the essence of what this entire book is all about—God interacting in the lives of His people, making Himself known to all via the DNA molecule—and beyond.

The majority of the images included are from Wikipedia and Government agency sites. Thank you all for your contributions and please accept my apologies if I have failed to acknowledge your authorship or to adequately reference your works.

Various parts of this book are copies of unpublished articles that I have written over the past years that have been modified and incorporated into several of the chapters. These articles have been the catalyst for the continued effort required to organize the theological and scientific content contained herein; to firmly establish the unique relationships between the two disciplines; and to be able to present them as a unified whole, clearly, and understandably.

Heartfelt thanks go to my wife, Caroline, for hanging in there for all those years while I hammered away at the computer. Long hours and many days have turned into months, and months into years, with seemingly very little progress. This definitely is not the most efficient procedure when writing a book, but I am so thrilled to have been able to complete it and now to present it to you.

Again, thank you, everyone.

Contents

Preface

I am writing this book for three unique groups of readers:
1) for babes in Christ, 2) for the mature in Christ, and 3) for
Jews and Gentiles without Christ. These three categories of
people comprise all that there are in the universe.

• For Babes in Christ
I am writing to those of you who are newborn into the family of God, in hopes of bolstering your faith in the living Lord
and to increase your confidence in His written and spoken
Word. And most importantly, I would admonish you to step up
*as newborn babes, desire the sincere milk of the word, that ye
may grow thereby* (1 Peter 2:2). The apostle Paul had much to
say about newborn Christians and of how they ought to hone-
in on their study habits: *I have fed you with milk, and not with
meat: for hitherto ye were not able to bear it, neither yet now
are ye able* (1 Corinthians 3:2).

• For the Mature in Christ
I am writing to those of you who have become mature by
your faithful study of the Scriptures, in hopes of explaining to
you about the tripartite divisions of a person; of how the spirit
of every human being expresses itself in the image of the soul;
and ultimately, in turn, of what is happening within the con-
fines of the human body. Our body and soul are inextricably
linked to our Creator, and it is He who acts throughout the
body communicating with the world around us. What you see
of the body with your eyes is the outward expression of the
soul; the manifestation of the inward, divine image.

But now, move upward and enjoy the manifold blessings
obtained when the strong meat of the Word of God is presented and studied, and then apply what you know to be true to

your own life. Put away the childish things that sufficed when you first became Christian and embrace now the strong meat: *When I was a child, I spake as a child, I understood as a child, I thought as a child: but when I became a man, I put away childish thing* (1 Corinthians 13:11).

• For Jews and Gentiles without Christ

God has blessed those who are listed among the first two groups with His grace and mercy. You know who you are. Unfortunately, I do not know for sure those of you who make-up this latter group, whether you are of the elect of God but currently in unbelief, or of the non-elect. God only knows. What I do know is that everyone in this group is currently without Christ and in need of hearing the Word of God unto salvation.

Unfortunately, I can be of no value or service to those who are, in the eyes of God, truly of the non-elect. I can only commit to the command of God to go out into the world and preach the Gospel of Christ to everyone. But, even here, I am faced with a major problem: I do not usually speak with a large majority of people that I don't know. So, what am I to do? Well, I wrote a book that you may begin to hear!

So I am writing this book to those of you who care little or nothing about the subject matter, in hopes of explaining to you the difference between you and me, as of the difference between night and day. For those who will hear, listen closely and give ample attention to what you are about to read, and know this, that assuredly:

Christianity is the highest form of religion and Christians constitute the highest form of humanity.

This book, then, is for everyone, whether they be babes in Christ who yearn to advance in their faith; whether they be mature in Christ who are wise in wisdom and understanding; or whether they be those who do not know Christ as their personal Savior but who wish to obtain a sincere desire to become babes in Christ. Most earnestly do I hope that in these pages may be found a rich source of blessings for everyone, regardless of position or status, and that they may convey fresh evidence to the mind as to the truths contained in His Word. May God grant to all a full measure of understanding.

Introduction

Are you ready to tread where none have ever gone before? Are you ready to knuckle-down and endeavor to identify and to appreciate those things that our awesome God has prepared for us? Let's face it. Scientists, theologians, teachers of the Word, lawyers, doctors, senators, presidents and potentates have all failed to recognize and to consider that when the Bible speaks to us, it really speaks to our biology! We are all biological creatures! We act and react to what excites our cellular DNA molecules! After all, we live in both a theological and a biological world, and somewhere down the line the two are bound to intersect and bind into one, unified whole.

Christians have always claimed that God has savingly regenerated their bodies to newness of life—to an incorruptible life—a life now lived without sin. But how did He do it? I look the same, I feel the same, and I eat the same, and yet I am *not* the same! How can this be? Let's drill down deeper into this statement to see just what's going on here.

There are four Biological Levels or phases of existence that we will consider which involves each of us (there are more than four). The upper level, where we are all accustomed to be living, is the Populational Level. This level is likened to an acre of land with fields of row-upon-row of bright red roses glistening in the noonday sun. In this level, Christians are seen as attending crowded revival meetings in stadiums filled to capacity; going to church every Sunday morning to hear the preaching of the Word of God; singing in the choir; attending Sunday school classes to increase our knowledge of Him; to determine what God has ordained for each of us. This is the level where we go about our daily business; when we get up in the morning to have our breakfast, to brush our teeth, and then off to school or to work. All day we are at it, and then we arrive at home for supper and some TV; and soon—off to bed at night we go, to rest up for another day of the same.

To our amazement, the Bible speaks and acts almost exclusively on this level. It tells us about sin and its affects upon humanity and of the only cure available for its eradication; about a Babe born in a manger a long time ago; of how He grew up; how He suffered on a Roman gibbet and died for sinners; and, most importantly, of His bodily resurrection. His life constitutes and incorporates all the sacred doctrines of the Church; from Predestination and Election, to Regeneration and Glorification, to name but a few. This is our life at the populational level and it affects all of us, corporately.

But how do these doctrines play-out at the Organismal Level of existence, one level *down* from the populational level? To understand this level, we need to consider, not the whole field of red roses, but of one solitary rose where each rose is likened to one of us, mimicking what we do in our personal lives. Though we belong to the level above us in a corporate manner, of everyone together without exception, this time it is in a singular, individual manner. It's like looking into a mirror, and what do you see there?—You yourself!

As the rose starts from a seed in the ground, springing up toward heaven at the drop of a rain drop, lifting its petals higher and higher toward the sun for nourishment, we as individuals lift our head toward heaven to attempt to understand the Word of our God; a word not unlike an instruction manual to offer help and guidance for our daily subsistence.

If you never observed yourself in an MRI image, you may not know that at this level you consist of organs, muscles, bones, skin, hair, and blood to name but a few. You still belong to the upper level but now you need to consider yourself at this level, regardless of what all the other Christians are doing in their lives. They have their own mirrors to peer into.

At this level theology starts to take-on meanings closer to the heart for now we start to seek answers to some direct questions. What think ye of Christ? Did He die for your sins? Rise from the dead? Coming again? Major doctrines now seem to be getting very personal, attaching themselves to our very own body; interpreting everything we do or say; going everywhere we go. These are good things, but notice the difference. Now the doctrines seem to start to take-on newer or added meanings as they get closer and closer to us, and start to permeate the skin. It's like looking at an ice cream cone—circular and wide at the top at the populational level, becoming narrower and more specific as we descend towards the bottom.

For instance, have you ever witnessed an open-heart surgery? Right away the question arises—when Christ died for you, personally, what organs did He actually die for? Did the Holy Spirit regenerate your entire body? Only your heart? Your kidneys? Your spleen? A little of everything? What?

As we descend further, we come to the Cellular Level, two levels down, where we notice those little critters floating around all over the place. We call them cells. An MRI won't do you much good at this level. They swim around because they are said to be alive, trillions of cells, all alive! At the populational level we are admonished to witness to a lost world. Here we are seen witnessing, not only to other people around us, but to ourselves! Hold on now, we'll attempt to explain this later. But, did you know that you are exceedingly more active and successful in communicating the Gospel at the cellular level than you are at the populational level?

We're going to stop now at the Molecular Level. At the start we saw row upon row of people worshipping God (Populational Level), including you and me; then zeroed-in on one person (Organismal Level), of how he consists of organ systems numbering less than one hundred; then lower in level yet but greater in number (Cellular Level) at 10^{14} cells; and now we arrive at a level so enormously complex, consisting of more body parts than all the sands of the sea (Molecular Level). Oh yes, lower levels do exist; the atomic level and lower yet into the sub-atomic level, where it is said that there may not be enough atoms in the universe to do the job.

Things are changing very rapidly. But know this: It is not our Christian theology that is changing—God forbid!—it is our scientific community that is changing; having advanced tremendously over the years, bringing to light the complex and intricate workings of the human body; that our forefathers had no knowledge of to incorporate into their writings and sermons. For centuries, the molecular and cellular levels of biology have remained hidden from our eyes; hence our theology has not advanced significantly into the lower levels as it should have.

All this leads us to the question of life: *Where* did life start? The molecular level consists of atoms and clusters of atoms called molecules, but atoms are *not alive!* If molecules are made-up of dead atoms then it seems reasonable to conclude that molecules also are not alive. But when we get to the cellular level and look into an electron microscope, we see molecular things that are *alive*, moving around in organized patterns. Not only that, but they are all *talking* to each other! What! Are you serious? Just what happened between the molecular and the cellular levels? Did the microscope inject life into the mix? I think not (and YES, they *are* talking to each other!).

The second question is equally mysterious: *How* did life start? We will not spend precious time considering life in general. We'll let the theologians and evolutionists argue through all that. What we will attempt to do is to elaborate on *how* the Holy Spirit *regenerates* a lost soul to newness of life, to life everlasting, and that starting at the lower molecular level and ending up at the populational level where we all now live.

The intent of this volume is to assist everyone who would attempt to read such a work, to transport the reader into his or her own body, to observe firsthand the marvelous and intricate world that lies within each of us. The inward parts of the human body have a story to tell, a fascinating tale of who we are as individuals, and of the invisible and mysterious interactions between our physical and spiritual worlds.

I will endeavor to present to you, as we travel through the various phases of our presentation, discussions of the observations of scenes witnessed with our own eyes; and to do so in an informative manner that will be interesting and understandable to those least initiated in technical and scientific jargon; information that would raise your level of spiritual and scientific understanding to heights unknown.

This book may have the beginnings of becoming a vehicle for a continuing debate, food for thought for academia theologians and scientists to ponder endlessly in Ivory Towers.

But now a word of caution, for some may find the subject matter contained within these pages to be troubling, for admittedly this is not your typical Christian storybook. It may be the catalyst for an uprising among pastors and theologians; for seminaries and denominations to expand their curriculums and basic doctrines; for professors to retool; and for scientists to stand amazed before our awesome God. For this, I am truly sorry. But as Martin Luther said many years ago, as he nailed his famous ninety-five theses on the door of Castle Church in Wittenberg, Germany: "Here I stand, I can do no other."

"The spirit of man is the candle of the LORD, searching all the inward parts of the belly" (Proverbs 20:27).

List of Illustrations

LIST OF ILLUSTRATIONS

PART I

Predestination, Caterpillars
& Thermodynamics

1

THE JOURNEY BEGINS

---

*"Man does as he wills, but God makes
him do as He wills"*
Charles Spurgeon

This is the story of a typical Christian: of *how* he got what he has, of how he will live his life *with* what he has, and, ultimately, of where he will spend eternity *because* of what he has. He is referred to in this context as the *Elect*.

This is also the story of a typical non-Christian; of how he will live his life as he sees fit, and, ultimately, of where he will spend eternity because of the way he has lived that life. He is referred to in this context as *unbelievers,* or, as the *non-Elect*.

The unbelieving, or non-elect person, is here likened to a *virtual* Cancer Patient (I say virtual and *not* actual), who is characterized as if suffering from the dreaded cancer disease. It is not the actual disease that is in view here, but it is the *manner of life* lived by every non-Christian person; mimicking exactly the expressions and ultimate results due to the effects of the actual disease, as manifested in his body.

The intent is to convey to you how our awesome God is at work behind the scenes, in how He manages the *inward,* intricate functioning of the human body. We start at the Molecular Level of life—the beginning of all things—then progress through the Cellular and Organismal Levels—advancing in order and complexity—and ultimately resurface at the Populational Level; expressing itself in *outward* appearances and actions that visually define a Christian in the world today.

So....What is happening to you within your body? What is it that makes you and me want to worship and pray to our God that gives us so much love, joy, and peace; but, at the same time, why do others feel no inclination to do the same? Why do Christians welcome death as a means whereby we may enter heaven and finally meet the Lord Jesus face to face and there to adore and enjoy Him forever? But, why do non-Christians seemingly welcome death as a means whereby they may enter hell and finally meet Satan face to face and there to loath and abhor him forever? The distinctions, dear readers, are dramatic and are as far and wide as the East is from the West.

Come with me, hold my hand as we explore together what makes you and me to be Christian, and trace the progress of when we first believed to that which we now enjoy in our position with Christ. To those of you who are non-Christian, currently in unbelief, (whether you are of the elect but have not yet been called-out, or whether you are of the reprobate, I know not which), I empathize with you, for at one time in my life I was as one of you, as one without peace and hope in the world. But, in God, I have found mercy and forgiveness!

Therefore, I extend to all, men and women alike, an invitation (as the very same invitation that I received a long time ago), to come and join with us in our search for the answers as to why we, as Christians, always feel the way we do, and why you, as non-Christians, can never feel the same way as we do.

THE BEAUTY AND MYSTERY OF LIFE

To understand the beauty and mystery of the Christian life, we first must drill deeper into the human body to view firsthand what is going on beneath the surface of the skin. The human body is a vast, intricate web of strange, seemingly non-living molecules going about their business without heralding their comings or their goings, making it practically impossible to gain an accurate understanding of the role each plays in the drama of life. This drama is so amazing, so perplexing, that it baffles the mind of all who would attempt to understand this strange world; a world not totally unlike our own; but of a world that is so infinitesimally minute, so infinitely fascinating, and so unbelievably complex. Nevertheless, this is where we are; this is where it all begins; this is where it all happens.

So, fill up the fuel tank and buckle up your seat belt. Turn on the hazard lights and hold on to your hat. Say good-bye to the kids! Ready? Say a short prayer and ask the good Lord for an abundance of wisdom, knowledge and understanding (you're going to need it!). Now, ponder the following in your mind— indelibly write these words upon the lintel and doorposts of your heart—that they may never be blotted out, never be forgotten, nor ever be denied:

• **Biologically**
Everyone has Cancer Genes, but not everyone is Cancerous. Everyone has Predestination Genes, but not everyone is Elect. Upon a divine fiat of God, a Protein Molecule attaches to the tail of a Gene and acts as a Gate-Keeper that turns the Gene *ON,* or turns the Gene *OFF.* It is *not* a random decision; you cannot *will* to be cancerous any more than you can *will* to be elect. If Cancer Genes are turned on, you will become a Cancer Patient; if Predestination Genes are turned on, you will become an Elect Christian. Cancer Genes when turned on *can* be turned off; Predestination Genes when turned on can *never* be turned off!

• **Theologically**
A Cancer Patient is here identified to be a *description* of the character and of the express *image* of the life led by the non-Elect, living without Christ, such as those who are *perceived* as suffering, but not actually, with the dreaded cancer disease. In this wise, Cancer *mimics* the life of the non-Elect; Joy—the life of the Elect. Cancer implies the Curse of Sin; Election—the Cure of Sin. Both Cancer and Election at first onset are non-experiential; you do not feel them, but they grow steadily. Cancer is Degeneration and Death; Election is Regeneration and Life. The non-Elect Cancer Patient experiences little or no relief: *for I bear in my body the marks of Satan.* The Elect Christian experiences profound joy and everlasting peace:

From henceforth let no man trouble me: for I bear in my body the marks of the Lord Jesus (Gal. 6:17).

GOOD GENES, BAD GENES
One of the greatest blessings afforded to man is to receive in the body *the marks of the Lord Jesus.* In Jesus' time, those who believed in Him were alluded to as His "slaves." They were branded, incised, or burned with "marks" (Gk. *stigma),* into

their flesh that identified the bearer as to whom he belonged. Instead of bearing the old mark of circumcision as was done in Old Testament times, now in New Testament times the body of all Christians bear the marks of His floggings and sufferings as recognition of His ownership:

> *We are troubled on every side, yet not distressed; we are perplexed, but not in despair; Persecuted, but not forsaken; cast down, but not destroyed; Always bearing about in the body the dying of the Lord Jesus* (2 Cor. 4:8–10).

It would not be unreasonable to categorize the human genome as an abundance of genes that we will refer to as *utilitarian,* responsible for *amoral* attributes. They possess and direct the main functionality of bodily growth, daily maintenance, and continual preservation that are applicable to all persons. We would also introduce the concept of a small subset of genes in the genome as *predestinarian;* that is, they are responsible for the *moral* attributes of all individuals, of the elect and of the non-elect (there are no other classifications of people); of their positions, attitudes, and modes of existence.

Each one of us is given by God an abundance of predestination genes embedded within chromosomes in the human genome. We will ultimately use but a few, since not all predestination genes will be turned on to produce an abundance of growth proteins of a particular *specificity.* Only those genes that God has selected for your life will be turned on, but those not directly associated with your life will never be turned on. Some genes will lay dormant in all individuals, some turned on, and others turned off. Which ones do you like? Which ones do you dislike? Which ones do you think are active in your life?—(God only knows!).

The very first mark that distinguishes a Christian is the mark of election placed upon him by God that occurred in eternity past, when time was not, in that the Triune God met in the council chambers of heaven to choose some out of all humanity to be in a state of *pre-disposition* in Christ; placing upon him the mark of election—the mark of Christ—and to pass by others, the non-elect, placing upon them the mark of reprobation—the mark of Satan. As hard as we may try to resolve this issue in our minds, the Bible is not silent on these facts; for it is plainly evident that there is a heaven and that there is a hell;

that there are angels and men in heaven, and that there are angels and men in hell. There is no denying this. Contemplate, therefore, wisely!

Predestination is a biblical concept that contains theological and scientific doctrines and theories of life that men have pondered over, and struggled with, for centuries. Election and reprobation happen in time, at the molecular level, when the gatekeeper activates particular predestination genes that are predisposed to ultimately result in either a newborn child of God, or in a child of Satan. So, let me leave some thoughts to take with you as you read further in your journey, thoughts and terminology that will later come to mind to assist you in formulating a biblical concept of the marks of a Christian.

In my particular case, the doctrines of hell overwhelmingly scared me. So for years I've looked to the Scriptures for guidance and strength that ultimately led me to a mindset of determination—so essential for a lasting relationship with the Lord. Therefore, for me and my house, we *will* serve the Lord:

> *And if it seem evil unto you to serve the LORD, choose you this day whom ye will serve; . . . but as for me and my house, we will serve the LORD* (Josh. 24:15).

FREE WILL OR NO?

In mentioning the terms "elect," "non-elect", "predestination", and "reprobation", do you believe that by using these specific terms it brings to mind that there are but two avenues that one can take to arrive at the gates of heaven; one way by being selected (or elected) by a Sovereign God, with no moral or religious input from the one selected (the *Frozen Chosen* as the Presbyterians would have it); and the other way by one using what is called his or her own free will to choose their own way to heaven (the *Free Willers* as the Baptists would have it)? Please be at ease, as both groups mentioned are classified as among the elect with no indication of how one would obtain that position. I am simply using terminology that God used through His chosen authors: *For whom he did foreknow, he also did predestinate* (Romans 8:29).

Now before you start sending emails to me asking of what persuasion I am, whether I believe in "free will" or in "free grace", I will truly answer that I am a "Five Point Calvinist" who believes in the total Sovereignty of God in Electing Grace. I firmly believe that God has foreordained whatsoever comes to pass; that He made all things of nothing by the power of His Word in the space of six literal, twenty-four hour days; that God preserves and governs all His creatures and all their actions. This is my theology. This is how I was brought up from my youth. Predestination is embedded in my nature, in my soul, and assuredly, as we shall see, in my *DNA molecules.*

Though I believe free will to be doctrinally in error, God will not allow His sovereign plan of redemption to be diluted by mere thoughts of misguided men and women; so I still love you as brothers and sisters in Christ and will forever enjoy your presence in glory.

Having said that, know this—that free will means there is no higher external authority that can exert an influence over the application of my will. It means to me exactly what it says; that if my will is absolutely free, then no human, animal, potentate, or king, no matter how influential they may be, can influence my decisions in the minutest degree (God excepted).

In the real world of mere mortals, God is seen to possess the ultimate free will, for He is the Governor of the Universe, the Creator of all. But—does God really possess free will?

When God wills, He need not apply to a *higher* authority as there is none higher. God is in a position of ultimate authority as He can will for anything in the universe and receive it without any objection from anyone. He could will for everyone to be saved, without exception, and everyone without exception will assuredly be saved, whether elect or not. Now, that's power. Why, He could even will for all the angels and people in heaven to go to hell for a few days. Who is it that could say no?

Needless to say, God *does not* possess free will! Write this in your memory bank—*God wills according to His nature!* His nature establishes His will. He would never will to create humanity and then will that everyone be in hell. He could do just that *if* His will was free, as we all could never enter an objection in a court of law. But, the Bible assures us that this is impossible, as free will is irrational and non-existent in God. His divine nature disallows Him to act capriciously.

Sad to say, humans also will according to their nature. Our nature ultimately determines our destinies, not our wills. The Bible repeatedly asserts that human nature has fallen into sin and can do nothing but sin, which can never please God. If you ever want to go to heaven, then God has to be the One to do it.

THE SOVEREIGNTY OF GOD

The Sovereignty of God is one of the major doctrines in the entire Scriptures, attested to, not only by each of its authors, but by God Himself. He is not silent when it comes to the doctrine of election and reprobation, especially when it also applies to those who have lost precious little children, primarily during pregnancy; soon after birth or later in life, but before the "age of accountability." These children are in a separate category for they have not had the opportunity or the privilege of understanding the written Word of God, let alone to apply those things that may ultimately determine their future.

As soon as one male sperm mates with one female egg, a new living human being has come into existence. In the sight of God, these are the "little children" as depicted in Matthew:

> *Then were there brought unto him little children, that he should put his hands on them, and pray: . . . But Jesus said, Suffer little children, and forbid them not, to come unto me: for of such is the kingdom of heaven. And he laid his hands on them* (Mt. 19:13–15).

When Jesus *lays His hands on little children,* you can be sure that they are redeemed by the shed blood of the Lamb of God, no matter if they have done good or evil during their short lifespan on this planet; they are of the elect and eternally secure in Him, basking in His presence for all time:

> *For the children being not yet born, neither having done any good or evil, that the purpose of God according to election might stand* (Rom. 9:11).

Moving forward, if you are still uncertain as to which way you obtained your salvation, once again please be at ease. It makes no difference with respect to your responsibility in the matter as long as you *confess* in your heart that Jesus Christ is Lord and believe that He died for your sins at Calvary. That is the definition of a Christian—no other affirmation is needed, that will come later in your spiritual walk.

But if you still don't know for sure—if you are living in sin; have never heard of the way to obtain salvation; or simply never cared one way or the other concerning Christ—be advised that it is a simple matter to change your feelings; a change from what you *believe* you are, of a non-elect status, to what you *actually* are, of an elect status; by simply doing what the elect have always done—*confess!* That's all that needs to be done! You do not have to prostrate yourself on the ground, pour ashes all over your head, chant as you go around in circles three times, or even fast for forty days and forty nights. The penalty for *all* your sins has already been paid! The debt you owed to God has been canceled! It's as though the doctor informed you that you have an incurable disease, cancer, but then said there is an instant cure available with no chance of a return of the disease. What could be better for you than that?

Listen! It is so easy for people like you to become Christian that it boggles the mind as to why everyone just doesn't do it. If you were to ask me, I'd rather be living my life as a Christian and be saved by the shed blood of the Lamb of God; to be in His presence and to enjoy life in Him for all eternity; than to be *thinking* that I am of the non-elect and be deluded by the lies of Satan; lost and separated from God, destined to suffer for all eternity. The choice seems to me to be a no-brainer, so simple, so reassuring, and yet so elusive to billions of people.

WATCH THOSE MOLECULES!

I also talked very briefly about molecules and genes that I hope would whet your appetite to ask: "What do all these molecules have to do with the elect and the non-elect?" Well, everything! The human body is an amazing collection of biological material that for centuries has been the object of scientific investigation, making important strides into the minute workings of the basic molecules of life.

You may not know this but when someone proclaims Jesus as Lord, at that *instant*, there is set in motion within the human body, changes that stagger the imagination of the mind. Men and women are mystified. Science is baffled. Your body takes on a dynamic force never before experienced; a force so powerful that it begins to change the very structure of every molecule in your body. Because of this divine intrusion, this intervention, you are forever changed and will inevitably take on a new mode of existence, a dynamic mode that has never been experienced before. It is actually a new life, a new creation, characterized as being *in Christ!*

> *Therefore if any man be in Christ, he is a new creature: old things are passed away; behold, all things are become new* (2 Cor. 5:17).

What is it that *old things are passed away?* And what is it that *all things are become new?* What are these old and new *things?* I mentioned this before, we are all biological creatures; we do nothing unless our biology permits us to act and behave in certain ways in response to internal or external stimuli.

It is our *old biology* that passes away and it is our *new biology* that becomes new. Make no mistake about it, have no doubts about it, the answer to these eternal questions of life and death, of final destinies, lies *within* each of us.

We are now at *war* within ourselves, and the *body* is the entire field of the battleground; a war of one type of flesh against another type of flesh; a war against the flesh that we were born with from the womb, and the flesh that we are striving to become. It is a battle of immense proportions, typified and wonderfully illustrated in the realm of nature by the lowly Caterpillar and its soul-mate, the beautiful Butterfly. Who can deny the immense struggles that occur within the caterpillar as it miraculously transforms into another creature; a virtual textbook in the insect kingdom designed for those who have eyes to see of what our Creator has accomplished in His elect.

CHAPTER THOUGHTS

I am writing this book, firstly, to those who are Christian, in hopes that you will find material that is exciting, revealing, and stimulating. May you gain a heighten appreciation for the Word of God—to study its contents more earnestly with respect to what you have gained in reading this volume and of the silent, inward changes that have occurred in your body; of the changes still going on; and of the outward appearances of those changes that are sure to follow.

Do not think for one moment that your molecules and genes do not know, or even care, about what has just taken place in your life. Even though your genes never went to kindergarten or have ever taken a three credit biology course in college, they seem to possess an intelligence that no Nobel Prize winner in the scientific field of Molecular Biology could ever obtain. Who among us can adequately explain the bodily changes occurring within a caterpillar that we can actually observe and learn from, a complete change into a new creature—a butterfly! Can anyone adequately explain all that?

Secondly, I wish to invite those of you who *act* as if you are non-elect but are not, because I desire for you to know the Christ that I know; to consider that becoming a Christian is not just a theory for scholars to debate; a fable for the sophisticated to ignore; or a riddle for skeptics to solve. But instead, it is the most important decision in the universe that directly relates to your future. Your life depends upon it!

I wish to describe to you a new understanding of the origin and progress of how a Christian comes about and how he is to live his life with what he has just acquired; with both tangible and intangible *marks* that distinguish him as such. It is not simply a variation of a traditional theological view but a truly new position that even I have not hitherto foreseen nor have previously read. It even boggles the mind as to how I could have written about all this myself, for I strongly believe that the Lord Himself must be involved in this endeavor, as my knowledge does not extend beyond the meager learning acquired during a lifetime.

But be sure of this: In the fullness of time, God sent His only Son that He may bestow His love upon those whom He has chosen, that election may stand true. God determined, even before you were born, that it is all of His sovereign will because that is His nature to do so. But for now, wait for the time when He will call you out, an *effectual calling* that will change your life forever.

Finally, I am writing to assist you to see how all these events take place within your body, both chronologically and qualitatively, and hopefully to realize when all is said and done, that I have been of some help in presenting to you a glimpse of Jesus Christ and Him crucified, my Lord and my Savior, whom I yearn to share with you. If you accept what has been shared with you thus far, and are not intimidated by my crude rhetoric in attempting to present some of the basic facts of life, then I encourage you to—*read on!*

2

LEAVES, WORMS, BUTTERFLIES
& *T. U. L. I. P. S.*

"There is nothing in a caterpillar that tells you
it's going to be a butterfly"
R. Buckminster Fuller

If you want to see something in nature that is all around us, that has been with us for centuries, that sure looks like a transformation from one kind of flesh into another kind of flesh, mimicking the regeneration of an elect person from being non-Christian to becoming Christian, then I invite you to explore with me the life of the lowly caterpillar, a prime example as found in nature and an excellent illustration of an elect Christian. Born into the world as a lowly grub, the caterpillar grows up into a voracious destroyer of every green leaf it can sink its teeth into, then silently hides in a cocoon and dies. However, is this the end of the caterpillar? Is this all there is to the life of an insect made by God? Are we to continually purchase insecticides to keep them off of our favorite flowering plants, or are there some lessons of life to be gleaned here.

This is a subject that is little understood and embraced even by those Christians who confess the great doctrines written in their creeds. The subject here is to seek out and to expound the truth of the Divine Sovereignty of God in Electing Grace as declared in the Word of God. The theology of the apostle Paul in Romans 1:16: *For I am not ashamed of the gospel of Christ,* should be the cry heard around the world.

The Truth of the Word of God must be reclaimed and vindicated from Error, for Truth must expose Error wherever it

may be found. My very salvation in Christ did not depend upon my *will* to accept Him but upon His mercy and grace to
elect me. Who is it that accuses Him of unrighteousness? God
is not unrighteous for not electing you, for you have no excuse
for not believing in Him. His nature is free of all human criticism, for mercy and compassion belong to God:

> *What shall we say then? Is there unrighteousness with*
> *God? God forbid. . . . I will have mercy on whom I will*
> *have mercy, and I will have compassion on whom I will*
> *have compassion. So then it is not of him that willeth,*
> *nor of him that runneth, but of God that sheweth mercy*
> (Rom. 9:14–16).

TRUTH AND ERROR MET

As one looks up into the starry expanse to gaze at sights
unknown, or onto the earth below to behold the vast array and
beauty of life, one cannot help but to be amazed at the power
and majesty of our awesome God. Take heed to the Natural
World for it is a book authored and illustrated by no mere ac-
cident but by Divine design.

The following excerpt by Virgil C. Mayes is illustrative of
Truth as found in nature, second only to the Revealed Truth of
God as contained in the Holy Scriptures, and is descriptive of
the human predicament. The passage is quoted here in its entirety, as I believe it to be excellent in describing the intricate
workings of predestination as found anywhere in nature, as
God has set before us a virtual video of His sovereign thoughts.

Consider now and behold the caterpillar and the butterfly:

> Truth and Error met one day by the side of a beautiful,
> placid lake in a land called Eden's Garden. They spied a
> bed of Tulips nearby, and as they watched the Worms,
> chewing away on the greenery while Butterflies harmless-
> ly sipped away at the nectar in the blooms of the Tulips,
> they began discussing an extremely difficult subject. Why
> do *Caterpillars* eat and destroy the leaves of the Tulips
> while *Butterflies* seem to enjoy only the nectar from the
> bloom? Truth began to explain the subject of *Transfor-*
> *mation*. This confused Error to such an extent that he
> changed the subject and suggested that they go for a swim.

Truth took off his beautiful robe and jumped into the water. Deceitful Error grabbed Truth's robe, put it on, and ran away wearing the beautiful Robe of Truth. When Truth discovered that Error had fled wearing his beautiful robe, he decided that Error's treachery must be exposed. Having nothing to put on but Error's dirty garments, Truth decided to remain "The Naked Truth." As it turned out, this was a "blessing in disguise" so to speak. Truth needs no cover of any sort.

THE NAKED TRUTH

Truth knew that he had nothing to conceal or hide and is perfectly willing to be looked upon as he is. This is not so with Error. He knows that, to be accepted, he must at least be partially covered with the Robe of Truth. Since that fateful day at the water's edge, Error has been "passing himself off" as Truth. Nowhere in the entire world has he been so readily accepted as in the realm of religion. Error continues to wear Truth's robe. However, Naked Truth and Bare Facts are twins. They pursue Error wherever he goes. They are constantly being weighed in the balances of human judgment.

Error continues his relentless efforts under numerous and sundry titles; but he always wears the ROBE of TRUTH. He convinces his hearers that WORMS, some worms at least, love nectar from the flowers of God. Truth, on the other hand, tells another story. He called a group of Butterflies together one day and requested that they explain how they came to be butterflies. The Butterflies consulted with each other for quite some time and elected one to be their Spokes-butterfly.

THE BARE FACTS

The Spokes-butterfly began to speak. He said: "Here are the Bare Facts. I was eating leaves one day as I had done all my conscious life. As I was eating, something happened INSIDE of me. It had *never* happened before. It has never happened since. I began to yawn. Webbing began coming out of my mouth. I cannot explain it. It *must* have been something our *Great Creator* did inside of me. I passed through a dark and awesome experience. The next thing I remember is that I was *hungry*. It is strange indeed. I was repulsed by the very thought of *bitter green leaves* but I still was hungry. There was a *desire* in me that I had *never* known. I was *wet* and *helpless*. A gentle breeze (I believe you call it *wind*) began to dry me out and before I knew what was happening to me, I discovered I was a *different creature* and something was *drawing* me to the *nectar* of the Tulips. You asked me to explain it. I cannot. The other *Butterflies* said "amen" and began to sing Amazing Grace. Not even one of them could remember wanting to sip nectar before they were transformed.[1]

How closely our spiritual birth follows in-step with that of the lowly caterpillar and the beautiful butterfly! The Christian is also about change, change of our natural body into a spiritual body, the very metamorphosis of our soul. As did the caterpillar, our body is gradually remolded into different functioning parts; we did not grow, we did not mature, we just changed—totally changed! Chemists, biologists, evolutionists, and theologians all retreat in confusion before this miraculous conversion process performed by our awesome God.

THE SIAMESE TWINS

The above quotation is so beautifully stated, yet so amazingly accurate that it baffles the mind of those who will not see or hear. Naked Truth and Bare Facts are seen as Siamese Twins that are inseparably joined together at the hip, and wherever Naked Truth goes, Bare Facts is sure to follow.

But not so with Error! Wearing the Robe of Truth, he skillfully maneuvers his audience to the position of wanting to go for a swim rather than to join in and sing Amazing Grace. Error must be exposed for what it is, for Error asks, as Pontius Pilate asked centuries ago: *What is Truth? (*John 18:38).

Error would have you to believe that Worms, the unregenerate of the human family, love to sip, as Butterflies do, the sweet nectar of the Word of God. But Truth asserts that, as Worms love to eat the bitter leaves of the Tulip, so the unregenerate love to eat the bitter leaves of sin. They love eating bitter leaves because it is in their nature to eat bitter leaves.

Regeneration happens when something strange occurs inside that is unexplainable, and which has never happened before. The caterpillar is seen here as dying and then being born-again, bearing the image of a butterfly; by analogy, the Christian is also seen here as *dying* and then being *born-again*, bearing the image of Christ.

THE FIVE-POINT PLAYERS

There are essentially two families in the world today, those of the non-elect, *The Deceitful Error Family*—portrayed by father Pelagius Error and his son, Arminius; and those of the elect, *The Naked Truth Family*—of father Augustine Truth and his son, Calvin. Calvin, influential in forming the Protestant Reformation of 1536 at the early age of twenty-four, started to write his masterful work: *The Institutes of the Christian Religion*. To sum up the distinctiveness of Calvinism, these truths are expressed by the popular acronym—*T.U.L.I.P.S.*:

- **T** – Total Depravity.
- **U** – Unconditional Election.
- **L** – Limited Atonement.
- **I** – Irresistible Grace.
- **P** – Perseverance of the Saints.
- **S** – Sovereignty of God.

THE FIVE POINTS CONTRASTED

Since the time of the Reformation, Arminianism has been the chief rival to Calvinism, tending to open the door to Liberal Theology. Arminius countered with his Five Points of Arminianism that must have a God that is not sovereign but can be manipulated and held in bondage, for He has been dethroned by Free Will. The Five Points of Calvinism, expounded by The Naked Truth Family, and the Five Points of Arminianism, expounded by The Deceitful Error Family, are briefly contrasted below to introduce the two systems of theology that divide the Christian church today, obtained from L. Boettner's work, *The Reformed Doctrine of Predestination*.[2]

1. **T** – **TOTAL DEPRAVITY**

Bare Facts

But if our gospel be hid, it is hid to them that are lost: In whom the god of this world hath blinded the minds of them which believe not, lest the light of the glorious gospel of Christ, who is the image of God, should shine unto them (2 Cor. 4:3–4).

Deceitful Error Family

"Although human nature was seriously affected by the fall, man has not been left in a state of total spiritual helplessness. God graciously enables every sinner to repent and believe, but He does so in such a manner as not to interfere with man's freedom. Each sinner possesses a free will, and his eternal destiny depends on how he uses it. Man's freedom consists of his ability to choose good over evil in spiritual matters; his will is not enslaved to his sinful nature."

Naked Truth Family

"Because of the fall, man is unable of himself to savingly believe the gospel. The sinner is dead, blind and deaf to the things of God; his heart is deceitful and desperately corrupt. His will is not free; it is in bondage to his evil nature, therefore he will not, indeed he cannot, choose good over evil in the spiritual realm."

2. U – UNCONDITIONAL ELECTION

Bare Facts

According as he hath chosen us in him before the foundation of the world, . . . Having predestinated us . . . according to the good pleasure of his will (Eph. 1:4–5).

Deceitful Error Family

"God's choice of certain individuals unto salvation before the foundation of the world was based upon His foreseeing that they would respond to His call. He selected only those whom He knew would of themselves freely believe the gospel."

Naked Truth Family

"God's choice of certain individuals unto salvation before the foundation of the world rested solely in His own sovereign will. His choice of particular sinners was not based on any foreseen response of obedience on their part, such as faith."

3. L – LIMITED ATONEMENT

Bare Facts

I have manifested thy name unto the men which thou gavest me out of the world: . . . I pray for them: I pray not for the world, but for them which thou hast given me; for they are thine (John 17:6, 9).

Deceitful Error Family

"Christ's redeeming work made it possible for everyone to be saved but did not actually secure the salvation of anyone. Although Christ died for all men and for every man without exception, only those who believe in Him are saved. His death enabled God to pardon sinners on the condition that they believe, but his Atonement did not actually put away sins."

Naked Truth Family

"Christ's redeeming work was intended to save the elect only and actually secured salvation for them. His death was a substitutionary endurance of the penalty of sin in the place of certain specified sinners. . . . The gift of faith is infallibly applied by the Spirit to all for whom Christ died, thereby guaranteeing their salvation."

4. I – IRRESISTIBLE GRACE

Bare Facts

A new heart also will I give you, and a new spirit will I put within you: and I will take away the stony heart out of your flesh, and I will give you an heart of flesh (Ezek. 36:26).

Deceitful Error Family

"The Spirit calls inwardly all those who are called outwardly by the gospel invitation; He does all that He can to bring every sinner to salvation. But inasmuch as man is free, he can successfully resist the Spirit's call. The Spirit cannot regenerate the sinner until he believes."

Naked Truth Family

"In addition to the outward general call to salvation, which is made to everyone who hears the gospel, the Holy Spirit extends to the elect a special inward call that inevitably brings them to salvation. The external call (which is made to all without distinction) can be, and often is rejected; whereas the internal call (which is made only to the elect) cannot be rejected; it always results in conversion."

5. P – PERSEVERANCE OF THE SAINTS

Bare Facts

My sheep hear my voice, and I know them, and they follow me: And I give unto them eternal life; and they shall never perish (John 10:27–28).

Deceitful Error Family

"Those who believe and are truly saved can lose their salvation by failing to keep up in their faith." (Arminians, however, have not been agreed on this point; some have held that believers are eternally secure in Christ, and that once a sinner is regenerated, he can never be lost).

Naked Truth Family

"All who are chosen by God, redeemed by Christ, and given faith by the Spirit are eternally saved. They are kept in faith by the power of Almighty God and thus persevere to the end."

CHAPTER THOUGHTS

Admittedly, the doctrines of Divine Sovereignty are *not* milk-diet doctrines. They cannot be fathomed completely by human reason alone, nor have they ever been completely understood by the godly men who have proclaimed them.

Whether or not one is willing to wear either name, does not alter the fact that he is either Calvinistic or Arminian in his views. Calvinism stands for the truth that salvation is of the Lord; Arminianism makes salvation the result of human merit. The one system postulates irresistible grace; the other postulates inherent human goodness.

Mayes refers to his writing as a "parable," a story in which a truth is illustrated and contains a moral lesson for its readers, an earthly story with a heavenly meaning. But a parable is often hid from the eyes of those who cannot or will not see. Here we see the caterpillar; he did not ask to be changed, he loved his life the way it was. His diet of bitter green leaves was just the right food to nourish his body in its present condition.

But then, something strange happened. One day he awoke and he was changed. He noticed that he was now a new and a different creature; a beautiful Butterfly that no longer loves bitter green leaves, but is irresistibly drawn to the sweet nectar of the Tulips. So it is with all those who love the Truth.

As R. Buckminster Fuller has here previously quoted: *"There is nothing in a caterpillar that tells you it's going to be a butterfly"*, so also: There is nothing in a person that tells you you're going to be a Christian!

God is Sovereign and He will do all that pleases Him. Stand tall, dear Christian friend, get a grip on it, and stop tiptoeing through the *T.U.L.I.P.S.!*

3

TWO INVIOLABLE LAWS

"The grass withereth, the flower fadeth: but the word of our
God shall stand for ever" (Isaiah 40:8)

The doctrine of predestination, as formulated in the history of the Christian Church, has been a constant cause of discussion and controversy, for many Christians have been unwilling to accept the doctrine in either form. To adequately address the statements expressed, those aspects of a person's moral and legal relationship to his God, more fully addressed hereafter, attention must first be made to several of the sciences in order to present sufficient information in defense of the arguments stated.

The worlds and all that is in them have been unalterably determined ages ago by the great scientific minds of our early scientists, to be continually refined during succeeding centuries. In the area of Physics, it is a well-established proposition that science stands or falls on but two of its cardinal laws, the First and the Second Laws of Thermodynamics, which are scientific, universal, and unalterable.

THERMODYNAMICS

Thermodynamics (from the Greek, *thermos,* meaning heat, and *dynamis,* power) is the study of systems that exhibit a temperature difference and involves the flow of energy from one location to another because of that difference. In essence, it is the study of energy flow, the rise and the fall of that mysterious quantity of which much is known, but of which no one has ever seen. The two Laws of Thermodynamics affect all systems in the universe, whether it is weather patterns, ocean currents, star formations, galaxy movements, or processes within the human body, none of which will ever escape its consequences and final results. The time rate of change of energy with respect to time is the *power* that drives the universe the way it is.

THE FIRST LAW AND LIFE

When God finished his creative acts of the universe: *In the beginning God created the heaven and the earth* (Genesis 1:1), and at the beginning of the seventh day before He rested: *And on the seventh day God ended his work which he had made; and he rested on the seventh day from all his work which he had made* (v. 2:2), and in order that man not become a god himself, to be able to perform creative works as God has done, He declared the First Law of Thermodynamics to be in effect; popularly referred to as the Law of Energy Conservation:

Energy can neither be created nor destroyed, but can be converted from one form into another.

In His infinite wisdom, God has determined that no matter what method man uses, he cannot create nor destroy a single atom. God has placed limitations upon man as to what he can do and what he cannot do. He can, for his many uses, convert the atom into pure energy, for energy and its sibling, power, drives everything that moves in the universe. Life cannot exist without the expenditure of vast amounts of energy. Energy may be used by every system imaginable but none of it will ever disappear from off the face of the universe.

Energy is mass (or matter) that can be seen, touched, or eaten, and resides in everything. The page that you are now reading, the book that you are holding in your hand, or the delicious fried catfish you are about to consume, is energy in the form of mass. You may call it a book, you may call it matter, but scientists refer to it as mass. Albert Einstein demonstrated that mass is but another form of energy, as expressed in his famous formula: $E = mc^2$ where E = energy, m = mass, and c = velocity of light squared.

Man has been given the privilege, the honor, and the responsibility for determining how to organize and use this powerful tool. But man cannot place limitations upon God as to what He can do and what He cannot do. God cannot be confined in a box. Scripture passages as: *There is no new thing under the sun* (Ecclesiastes 1:9) refers to, and is applied to, man, but not to God. God is sovereign and does as He wills. If it pleases Him to create another universe such as our own, then so be it. Who is man to say that He cannot?

THE SECOND LAW AND DEATH

But after Adam sinned in the Garden of Eden and death was pronounced upon mankind: *But of the tree of the knowledge of good and evil, thou shalt not eat of it: for in the day that thou eatest thereof thou shalt surely die* (Genesis 2:17), it was only then that God initiated the Second Law of Thermodynamics, popularly referred to as the Law of Increased Entropy. Because of the sin of Adam, and that sin conveyed upon all men, (Total Depravity, we saw this before), Entropy was introduced into the affairs of the universe and death became a new thing in the lives of men.

Entropy is the measure of molecular *disorder* and is the amount of *unavailable* energy in a system. An increase in entropy is not a good thing. When entropy *increases*, the net amount of available energy useful for work *decreases*: iron rusts; food spoils; stars burn out; people die—are daily instances of increase in entropy and decrease in complexity and order. It is a universal law and certain a law as exists in nature, a directional law that dissipates heat energy into space and is then lost. This world will not continue endlessly; the universe is constantly becoming more and more disorderly.

There will not be enough available energy in the future to do any useful work to sustain life. All the hotter, higher-energy reservoirs will have poured their energy into the colder regions of the universe and everything, everywhere, will be at a uniform temperature. In such a state, it would be impossible to build any more heat engines that could do any useful work—no more beautiful cars, no more luxurious homes, no more classy clothes—the universe would grind to a halt and eventually "die." The Second law predicts that any system that is left to itself will go toward greater disorder and decay.

In the final analysis, when all the available energy in the universe has been used for useful work, and the residual changed to random heat energy, consisting of random molecular motion and everywhere exhibiting the same uniform temperature, is referred to in scientific literature as the *Heat*

Death. In this state, all activity stops! In the game of chance, the Second Law is likened to the house dealer and he deals us a hand—you simply cannot get out of the game, *you must play*:

> In the great game of the universe, we not only cannot win, we cannot even break even.[1]

Cancer and its effect upon the human body are but the result of the Second Law in action. Whether it is cancer, old age, or some other dreadful disease, we are all subject to it and no one will ever escape. The prophet Isaiah spoke of the Second Law as being implemented by God, as to touching both the natural order in general, and to humanity in particular:

> *The voice said, Cry. And he said, What shall I cry? All flesh is grass, and all the goodliness thereof is as the flower of the field: The grass withereth, the flower fadeth: because the spirit of the LORD bloweth upon it: surely the people are grass. The grass withereth, the flower fadeth: but the word of our God shall stand for ever* (Isa. 40:6–8).

Heaven and earth are transitory. Both are compared to an old garment that one day will be rolled up and discarded in the big trash heap of the universe. The current heaven and earth will pass away with everything in it: *the elements shall melt with fervent heat, the earth also* (2 Peter 3:10). Both are now in a state of decay. All creation is transitory and locked in a downward spiral of deterioration and eventual death.

The Second Law of Thermodynamics cannot naturally be suspended or reversed, for that which God has ordained, man cannot annul. Rust can never turn back into beautiful malleable iron; a pile of decayed rot can never turn back into a delicious morsel of food; burned out stars can never relight; and dead men can never come back to life—for the Lord Himself has said: *For the Lord of hosts hath purposed, and who shall disannul it? and his hand is stretched out, and who shall turn it back?* (Isaiah 14:27).

Is there no relief? Is there no escape from that which God has ordained? Surely, man is capable of solving this dire dilemma—the world's most perplexing problem.

ENTROPY AND HUMAN NATURE

When Adam sinned in the Garden of Eden, he underwent a drastic change in his nature; he was now able to sin at will: *As by one man sin entered into the world, and death by sin; and so death passed upon all men, for all have sinned* (Romans 5:12).

Something happened in man's genetic makeup to cause him to eventually die. *Entropy in his genome increased!* Information in his genetic structure decreased and has been forever lost, *because the spirit of the LORD bloweth upon it* (Isaiah 40:7). He cannot now ward-off diseases as he was designed to do *before* the Fall, and death is now inevitable. That is the Curse and it continually plays-out in the human body.

All science bows before the Second Law, and apart from a higher power resident outside of the universe, everything left to itself will ultimately go to pieces. The Second Law of Thermodynamics is seen as never to be overthrown....(Or will it?).

But for every Curse, there is a Cure! Jesus Christ died on the Cross at Calvary, the One *which taketh away the sin of the world* (John 1:29). An acceptance of faith and trust in the atoning death of the Lord Jesus inevitably results in a new nature within the human body. Man now has *two natures*, an old nature resulting after sin has come, and a new nature resulting after faith has come.

CLOSED OR OPEN?

In a closed or isolated system, a system considered with no external sources of both energy and information impinging upon the system, processes have no alternative but to go from a high degree of order and complexity to a lower degree of order and complexity, which is entropy in action in the Second Law. But are there any truly closed or isolated systems? Is the earth, as a system, closed or open? All processes in the earth do follow the Second Law, decreasing in both complexity and order, but at the same time all sorts of plants are seen springing up in majesty and beauty, defying the Second Law!

Some scientists have concluded that no finite physical system on earth can be considered closed, since the Sun is an external source of energy pouring out huge amounts of energy into the earth to support useful work. The earth therefore cannot be truly classified as a closed system as there would be no growth, and that is totally against our observations. To initiate plant growth or any kind of *upward, complex organization* in any system requires *external energy, external information, and an external conversion process.*

As an example, imagine a meandering bull running amuck throughout a beautiful and eloquent China Shop. He will expend an awful amount of energy, but he will not extend the smallest degree of loving-kindness to a very fragile, beautifully designed and decorated ceramic tea pot. Energy needs to have a guiding influence, a degree of organization, an amount of conceptual information inherent to order its directions. Otherwise, raw energy is, in every situation, a destructive force.

While the First Law says that the total amount of energy in the universe remains constant, the Second Law essentially says that "I will divide your energy into two categories: 1) into useful energy, and 2) into useless energy." Now we have no problem with useful energy as we see it occurring throughout our daily lives—driving our automobiles, cooking our food, washing the dishes, and drying our clothes to name but a few. But the problem occurs when we see the gas tank emptying, our ovens won't light-up anymore, and the hot water heater is getting colder. Where did all the energy go to drive these useful systems?

Well, say—bye bye! You will never get it back again to perform the same work. No more energy to furnish the power to turn the wheels, the burner switch, or the agitators. No more information to plan vacation trips, to savor warm food, or to ever wear clean clothes again. We have finally reached the lower-end temperature point—absolute zero! WOW! Is it cold!

But God, in His infinite wisdom, will not allow His creation to reach such a point of disorder, to the point of total disarray. He has thus instituted the following criteria in order to preserve His creation, to *reverse or suspend the actions of the Second Law!* By this, a system must be able to receive freely the following three, primary and necessary conditions:

• **An External Source of Empowering Energy**
Of an extent within the system of sufficient magnitude, as in: *In the beginning God created the heavens and the earth* (Genesis 1:1), or as in: *And God formed man of the dust of the ground* (2:7)—Energy expended!

- **An External Source of Conceptual Information**
Of clarity of scope to organize each phase of complexity, as in: *And God said, Let there be light: and there was light* (1:3), or as in: *And breathed into his nostrils the breath of life* (2:7)—Information expended!

- **An External Source of Conversion**
Of exact dimensions to cause a transformation of scope, as in: *And the Spirit of God moved upon the face of the waters* (1:2), or as in: *And man became a living soul* (2:7)—Metamorphosis completed!

If the source of information is of a different order than that intended, then no amount of external energy to the system will suffice, and the endeavor will continue to decay according to the Second Law of increased entropy.

Considering the human body as an open system, Cancer Genes exhibit no source of empowering energy, a degenerative informational program, and no inherent conversion process to increase the order and complexity of infected cells, affording no cure. Both Cancer and Sin succumb to the Second Law.

In contrast, Predestination Genes do exhibit an external source of empowering energy to drive and to sustain the system from its previous form; an informational program of conceptual thought to order and to identify the means and direction for the system; and contains a regenerative conversion process to increase the order and complexity of infected cells, leading to a Cure for Sin. We firmly proclaim that *election suspends the Second Law,* to afford a new life that far surpasses that which Adam enjoyed before the Fall.

DIVINE MIRACLES

Divine miracles, acts of God not occurring spontaneously in nature, require that the Second Law be *suspended* during the time of the initial application. The Second Law, however, is still in full force, but act only in a subset to the Law. The Law itself is inviolable, and in the final analysis, will prevail.

MIRACLES AND JESUS
Grass and flowers spring up in life while the sun shines, adorned in radiant beauty and majesty, but take away the

heat from a blade of grass, the instructions for growth from the bloom of a flower, or the laws of nature from a kernel of corn, and in the final analysis—the grass *will* wither and the flower *will* fade. The external sources must be maintained.

So, what is a miracle? Notice three elements: 1) miracles are the supernatural works of Deity; 2) they suspend natural laws; and 3) they reveal the nature of God to a lost world. The apostle Peter spoke of Jesus as *a man approved of God among you by miracles and wonders and signs* (Acts 2:22). Miracles are therefore an integral part of Christ. He performed miracles and continues to do so. Miracles are in his nature to do. It is impossible that a miracle could have taken place without God's special intervention. They would simply be amazing events.

The resurrection of Jesus Christ is a prime example of a divine miracle, the idea that a Person, as He was, could be restored to life after being dead in the grave for three days. His resurrection is the best-attested event in all of history, confirmed by many eyewitnesses as recorded in Scripture, and in many Christian and secular sources. Jesus arose from the dead and this fact can be verified and investigated by the same criteria used to verify other historical events.

For a miracle of this magnitude to have occurred, a magnitude no less than that of the creation of the world itself, enormous amounts of empowering energy must have been supplied to reverse or suspend the Second Law for a duration long enough to offset decay in the grave!

I BELIEVE IN MIRACLES

I believe in miracles for the Bible tells me so, and what the Bible says, in my humble estimation, is true, for without miracles Christianity could not exist. Without the suspension of the Second Law, without enormous amounts of energy poured out into the system, Christ could not have risen from the dead. A change of this magnitude in the earth, as in a rising from the dead, demands a suspension, and the suspension demands an Intelligent Creator who purposefully created the universe, got it running, and now continually maintains its existence.

I believe in miracles because of the sinless life of the only Person that ever lived that could truly redeem humanity of the sin nature and its lethal effects; the only way to defeat the sickness and be free of its presence. Although I have never witnessed a miracle on the order of a resurrection, or of changing water into wine, I believe in miracles because of:

- **The Virgin**
Behold, a virgin shall conceive, and bear a son, and shall call his name Immanuel (Isa. 7:14).

- **His virgin birth**
To a virgin espoused to a man whose name was Joseph . . . and the virgin's name was Mary. . . . And, behold, thou shalt conceive in thy womb, and bring forth a son, and shalt call his name JESUS (Luke 1:27, 31).

- **His walking on water**
And in the fourth watch of the night Jesus went unto them, walking on the sea (Matt. 14:25).

- **His casting out Demons**
Let us alone; what have we to do with thee, thou Jesus of Nazareth? art thou come to destroy us? I know thee who thou art, the Holy One of God. . . . Hold thy peace, and come out of him (Mark 1:24–25).

- **His transforming water into wine**
Jesus said unto them, Fill the waterpots with water. . . . When the ruler of the house had tasted the water that was made wine, and knew not whence it was: (but the servants which drew the water knew) (John 2:7, 9).

- **His raising of the dead**
And when he thus had spoken, he cried with a loud voice, Lazarus, come forth (John 11:43).

- **His burial and resurrection**
And that he was buried, and that he rose again the third day according to the scriptures (1 Cor. 15:4).

- **His second coming**
For the Lord himself shall descend from heaven with a shout, with the voice of the archangel, and with the trump of God: and the dead in Christ shall rise first (1 Th. 4:16).

In light of all this, what then is the *source of empowering energy* to furnish the power to enable the Second Law to permit a divine miracle, such as to transform a non-Christian into a Christian; and what is the *source of conceptual information* that instructs existing biological molecules how to organize themselves into new hearts, livers, and kidneys; and what is the *source of complex conversion* that empowers the new system to function exactly as it was designed to function?

CHAPTER THOUGHTS

I have witnessed a modern-day miracle with my very own eyes every time a lowly caterpillar morphs into a beautiful butterfly that no one knows, nor ever will know, of how he did it. He did not want to do it; he did not know how to do it; he just did it. The life of a caterpillar serves as a prime example of the metamorphosis of a Christian via the actions of predestination and election.

I define a *micro miracle* as a miracle that produces a micro creation that takes place in a *horizontal* direction, a variation in an existing species, as in the cat family—a lion and a kitten. These two animals are different and show horizontal variation, but belong to the same Cat family. I also define a *macro miracle* as a miracle that produces a macro creation that takes place in a *vertical* direction, not just a horizontal variation in a species, but itself a *new species*, as the transformation of a caterpillar to a butterfly in the insect family.

The definition of species is different for many scientists and is not a standardized term. Many scientists define species as a population of life forms that are reproductively isolated from other similar species. I wholeheartedly agree with this definition: for butterflies breed with butterflies; caterpillars do not intermingle with butterflies; Caterpillars migrate as far as one flowering plant is to another flowering plant in your home garden; butterflies migrate thousands of miles. Caterpillars are seen to be very distinct from butterflies. Therefore, we claim that butterflies may be considered a new species, other than that of caterpillar (not everyone, however, will agree to this vertical mobility to a higher order and complexity, resulting in a new species).

But what are the specific inputs that would enable an existing human species to completely mutate to a higher-ordered species? First, an external source of empowering energy capable of reversing the Second Law of Thermodynamics, that enables the addressing of the DNA structure to make the necessary changes in nucleotide sequences that is different from all other DNA sequences; second, a conceptual information program that guides proteins how to organize themselves into hearts, kidneys, and livers of the new species; and lastly, a converter process to empower or to force non-living systems to become living. Now, that's a miracle!

TIDBITS OF PART I

- Everyone has Predestination Genes.
- Everyone has Cancer Genes.
- Gate Keeper turns genes ON/Off.
- Cancer mimics the life of the non-Elect.
- Predestination genes turned On:
    o The Elect;
        ♦ A Christian.
- Cancer Genes turned On:
    o The Non-Elect;
        ♦ A non-Christian.
- Cancer:
    o The Curse of Sin;
        ♦ Degeneration and Death.
- Election:
    o The Cure of Sin;
        ♦ Regeneration and Life.
- *If any man be in Christ, he is a new creature:*
    o The butterfly and the Christian, by metamorphosis.
    o The caterpillar—Nature's video textbook on Election.
    o The butterfly—Loves *T.U.L.I.P.S.:*
        ♦ T – Total Depravity.
        ♦ U – Unconditional Election.
        ♦ L – Limited Atonement.
        ♦ I – Irresistible Grace.
        ♦ P – Perseverance of the Saints.
        ♦ S – Sovereignty of God.
- How I love the sweet nectar of the bloom!
- How I hate the bitter leaves of sin!
- Christians!
    o Emulate a butterfly!
    o Love the Word of Life!

- The Naked Truth family loves the truth:
  - o Naked Truth and Bare Facts are Siamese twins.
- The Deceitful Error family loves error.
- First Law of Thermodynamics:
  - o Conserves energy;
    - ♦ Life goes on.
- Second Law of Thermodynamics:
  - o Entropy increases;
    - ♦ Everything dies.
    - ♦ Iron rusts.
    - ♦ Food rots.
    - ♦ Stars burn out.
- Boy, do we need some kind of miracle to save us!
  - o How?
    - ♦ Suspend the Second Law!
- Miracles:
  - o It takes Three to concur.
  - o External Source of Empowering Energy;
    - ♦ To *initiate*.
  - o Conceptual Information Program;
    - ♦ To *inform*.
  - o Converter Process;
    - ♦ To *activate*.
- I believe in miracles! Do you?
- Divine miracles, that is—
  - o Like His virgin birth!
  - o Like His walking on water!
  - o Like His casting out demons!
  - o Like His burial and resurrection!
  - o Like His coming again!
  - o Did I mention;
    - ♦ "Like His raising the dead?"

But for the greatest of all miracles that stand before our very eyes, come with me now, and let's take a closer look within the human body, to gaze upon the plethora of divine miracles that suspend or reverse the Second Law, that will surely baffle and stagger the human mind.

PART II

Microbiology 101

4

IN THE BEGINNING—*DNA!*

*"Lift up your eyes on high, and behold who hath
created these things"* (Isaiah 40:26)

C hristians have a flower by which to identify themselves;
it's called a *TULIP* and its nectar is as sweet as the
Word of God. The leaves on the stem of the Tulip are to
us *anathema*, bitter and unpalatable, but to caterpillars and to
those who love to sin, they are a morsel of delicious delight.

We have previously demonstrated that God creates according to the good pleasure of His sovereign will, and does so as
no mere mortal can. He is not subject to, or under bondage of,
either the First or Second Law of Thermodynamics, but uses
and manipulates them for His own purposes. God clearly indicates that He is continually showing Himself in His divine nature by all the things He has made and by providing all the
necessities of life. The very laws of the universe itself have
been tailored and fine-tuned for our use, comfort, and enjoyment.

For those who have eyes to see and ears to hear, you are
without excuse if you remain in unbelief, for God is continually
showing Himself as Creator, *upholding all things by the word
of his power* (Hebrews 1:3). He simply speaks the word and it
is done. Words are knowledge; knowledge is information; and
information is the driving force behind His power to do and to
create. From the beginning of the world, things that have always been visible and clearly seen are now becoming known,
as science discovers and learns more and more about the DNA
molecule, the Deoxyribonucleic Acid molecule, the molecule of
all living things.

THE MOLECULAR LEVEL

Like many of my generation, I managed to make it through my high school days without ever hearing of the term DNA. I was totally absorbed in sports and other things and not much interested in learning, let alone learning about those little things crawling around in the biology lab.

So now here I am today, well over a generation later, talking about a subject that I never studied, never thought about studying, and never thought I would someday ever begin to study. But here I am, talking about those things that just fascinate my mind to a degree that not many professional scientists in the field of biology would embrace or agree with what is to follow. But I am persuaded, that a Christian could never be Christian without those little things doing what they do, going around making what they make, and after all that, telling you all about who you really are and what makes you do the things that you always like to do.

With these declarations in mind, we need to look no further than to the human body for some of the most marvelous and complex activities in all of creation, and that through the lens of Molecular and Cellular Biology. We will observe firsthand the vast array of changes that permeate our bodies, to gaze upon and to marvel at all the biological activity performed in synchronous motion that, in all probability, should be classified as non-living molecules. Life is nothing less than a miracle in the realm of human activity, but is nothing less than, or more than, a common occurrence in the divine will and creative acts of our awesome God.

THE BUILDING BLOCK

The atom is the basic building block of all matter, a particle small enough for more than one trillion of them to fit on the period at the end of this sentence. Atoms are found in nature as pure elements such as the oxygen we breathe, or as compounds such as the combination of sodium and chlorine, the table salt that we put on our food for taste. We write the molecular formula of table salt as $NaCl$ to indicate that it has one atom of sodium, Na, and one atom of chlorine, Cl. Some molecules contain a few atoms but others contain large numbers of atoms and are referred to as macromolecules. Proteins of life are macromolecules, often containing tens or hundreds of molecules.

Living biological molecules are a symphony of billions of atoms all miraculously timed and coordinated to perform all the functions necessary for life. Amazingly, this symphony of life has only a few major players, only five required atoms: carbon (C), hydrogen (H), oxygen (O), nitrogen (N), and phosphorus (P).

THE DNA MOLECULE

One of the most important discoveries of the twentieth century in molecular biology was the discovery of the DNA molecule. It is not unreasonable to say that the DNA molecule is one of the greatest scientific discoveries of all time, as it answers to some of the most baffling questions about the origin and functionality of life itself.

DNA is the famous molecule of genetics that establishes each organism's physical and thus its inheritance characteristics. Each of us starts out in life with a DNA molecule much smaller than a dot on this page and within that microscopic sphere is six feet of molecules all coiled up in a little ball. Within the DNA itself is contained the entire code of life for each and every individual; the entire set of *blueprints* of who you are to become, of all your organs, of all your features, and of all your children. It is akin to a microscopic computer with a huge memory bank that stores vast amounts of information, and at the proper time and place, issues orders to various parts of the body to commence building cellular structures such as livers, hearts, and kidneys. Your DNA is like your thumbprint. It is yours and yours alone. Unless you have an identical twin, no one else on the planet has exactly the same DNA signature as you have.

DNA is a double-stranded molecule that is twisted into a helical structure, like a spiral staircase. Each of the two, six feet long rails is comprised of a sugar-phosphate backbone and numerous base chemicals attached in pairs, held together by hydrogen bonding. The four base molecules that make up the stairs in the spiraling staircase, composed of *amino acids*, are Adenine (A), Thymine (T), Cytosine (C), and Guanine (G), where A bonds with T, and C bonds with G, always. No other possibilities are permitted (unless you want to be abnormal!).

These base pairs are arranged in particular sequences and contain the codes necessary for life. They act as the letters in the genetic alphabet; combining into complex sequences to form the words, sentences, and paragraphs that act as instructions to guide the formation and functionality of the human cell. Figure 4.1[1] illustrates a section of the DNA molecule, showing the two side rails of the ladder and the base pairs of A bonded with T, and C bonded with G, in some particular sequence. Also, looking from either end of the molecule, along the horizontal axis, notice the right-hand twist of the side rails.

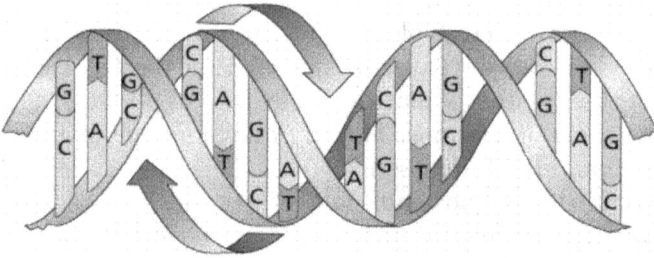

Figure 4.1: DNA Molecule

Drilling down deeper, Figure 4.2[2] is an excellent illustration of both side rails of the DNA molecule that precisely orients each of the components associated with the molecule along with their chemical formulas. This is the basic unit of life. This is what we are all made of. This is what we all look like, and this is but the first letter in the ongoing code sequence that determines our destinies. There is no other way and no other solution to life, whether human, animal, plant, or bacterial. All have the same or similar DNA structures and amino acid nucleotides, but yet we are all so very different!

The basic construction of the DNA helix staircase molecule consist of two side rails, or backbones, with bases connected to each other that link both rails together. Each rail is made-up of three subunits, called a *nucleotide*: 1) a pentose sugar molecule, 2) a phosphate group, and 3) one of the four nitrogenous bases, A, C, T, or G, describing one extremely small segment of the DNA molecule. To construct the total molecule, keep adding the above subunits together, the phosphate group of each nucleotide linked to the sugar of the adjacent nucleotide, ultimately forming the 6-foot long helical chain.

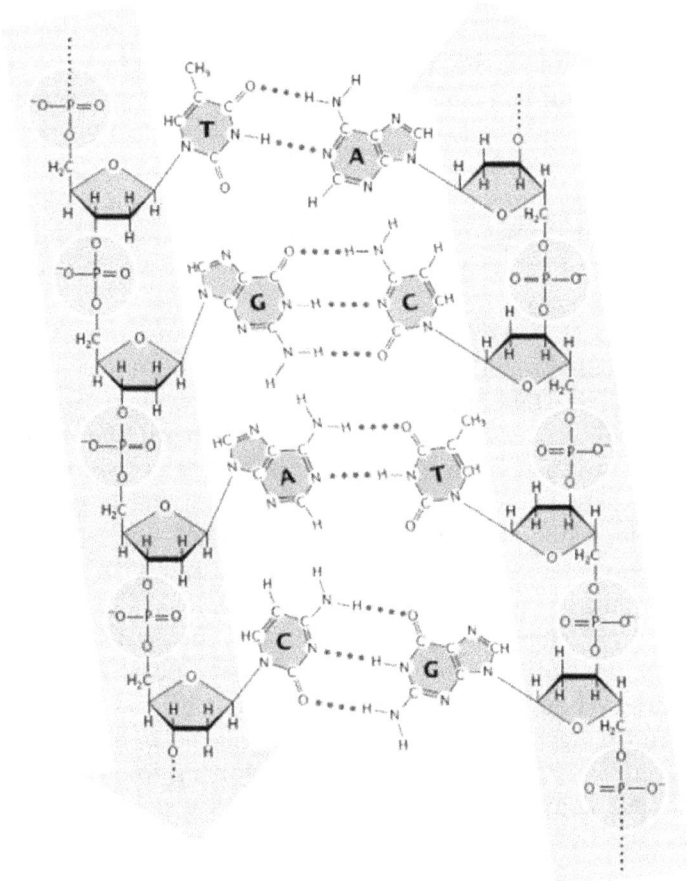

Figure 4.2: Genes Are Us

Plainly visible are the five-sided sugar molecules consisting of atomic elements carbon, hydrogen and oxygen, chemical formula $C_5H_{10}O_4$; the phosphate molecules PO_4 with phosphorus and oxygen; that together with the sugar molecules make up the side rails of the pentose sugar DNA molecule. The addition of the four nitrogenous bases, $C_5H_5N_5$ for adenine (A), that connect the two rails, comprise the basic segment of the DNA molecule, called a *nucleotide*. Also, notice where and how

all the components connect to each other in a specific order
that is identical in all DNA molecules with absolutely no variation. This is life at its basic core and the code of life is contained in the base pair sequence TGAC in the left rail paired
with ACTG in the right rail for the particular sequence shown.

The stability of DNA is primarily dependent on the strong
bonds that connect the atoms of their linear backbones. In addition to strong bonds, a number of weak hydrogen bonds are
important in interactions between these molecules and between groups within a single molecule. Unlike strong bonds,
which require considerable energy input to break them, weak
bonds are constantly being made and broken. As a result, they
readily permit molecular interactions, which are essential for
biological activity.

The length of DNA is expressed in the number of nucleotides connected together that comprises the chromosomal DNA
molecule. The DNA molecule is so complex that man, with all
his superior knowledge and intellect, with all the expensive
and intricate instruments in the laboratory that are at his disposal for experimentation, cannot manufacturer a single amino acid, let alone construct a DNA molecule!

Given this 4-bit simple code, TGAC paired with ACTG, can
you recognize yourself? This is what you look like at the molecular level; consisting of simple atoms, molecules, amino acids, and nucleotides; connected in a very special and unique
sequence. When you look in a mirror what do you see. You see
flesh and bones, eyes and ears, hair and teeth, but way down
deep there is a whole new world just waiting to be explored, a
world that contains not only life as we know it, but possibly
answers to that age-old question of—what is life?

DESIGNER GENES

Every thought that originates in the brain is directly related to your genes and ultimately to the way you live. If the
genes are not programmed to accept your choices, you will not
achieve what you are thinking. So put the mirror away, this is
what everyone looks like—get used to it! This is what makes
you do what you do and how you do it; and what I do and how
I do it. We will be talking about these pesky little things more
and more as we continue on our journey.

Each gene consists of a variable length of discrete DNA
molecules and may contain many thousands depending upon

the code contained in the nucleotide sequence. There are approximately 35,000 genes in the human DNA molecule, comprised of approximately 3 billion (3 x 10^9) nucleotide base pairs arranged in precise sequences. In Figure 4.3^3, each gene has only one job to do; that is to remember exactly how to construct a single protein molecule.

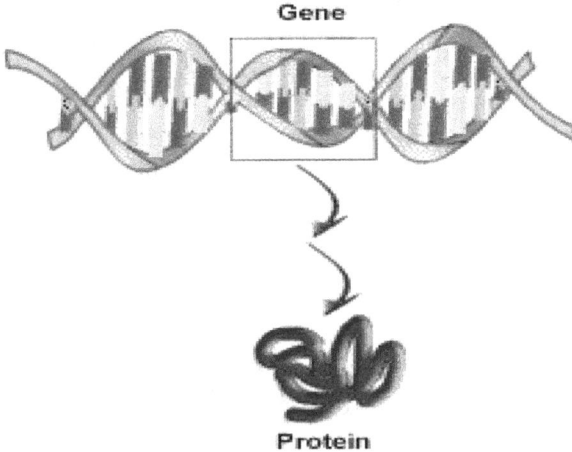

Gene

Protein

Figure 4.3: From Gene to Protein

THE SENTENCES

All living organisms are made up of cells. Skin cells are the cells that form the environmentally controlled enclosure that houses the variety of other types of cells. Inside each cell is a nucleus and inside each nucleus, among other complicated things, are 46 chromosomes, and inside each chromosome are DNA molecules, and inside each DNA molecule are many genes that are attached to the molecule like beads on a chain.

A gene in the DNA molecule is ultimately expressed into a particular protein. The A, C, G, and T nucleotide base letters of the gene act similar to the "0" and "1" alphabet in binary code as found in computer machine language, and communicate information to the organic cell by a lengthy process. As the carrier of information, the molecule is 45 trillion times more efficient than the computer's silicon mega-chip.

The structure of the DNA molecule is organized into *stretches:* stretches of genes whose function is to coil the DNA into chromosomes; stretches that "turn a gene on" and stretches that "turn a gene off"; and larger stretches whose purpose is not yet fully known to scientists, commonly alluded to as Junk DNA (jDNA). The molecule contains enough information to fill all the books in the world's largest libraries.

The gene sequence is the particular side-by-side arrangement of base pairs along the DNA helix strand. This order spells out the exact instructions required to create a particular organism with its own unique traits. The molecule acts as a template or pattern for the making of specific protein molecules that control the growth and activity of the cells, that in turn control the growth and activity of the whole organism.

The genetic code as found within a gene is vital to life as we know it. The code's chemical instructions are copied faithfully, time after time, and produce the same organism or Biblical *kind*, time after time; bluebirds produce only bluebirds, buttercups produce only buttercups, and humans produce only humans. Life now seems to be embedded in the nucleotide code.

Although the genetic code is remarkably complex, it is the information translation system inherent in the code that really baffles science. The base letters and words of the code mean nothing outside the language convention used to give those letters and words specific meaning. The words and letters printed on this page mean nothing unless the human brain translates the code into meaningful information and ideas. The chemical building blocks have nothing to do with the *origin* of the complex message! This is modern Information Theory at its core. We now understand that organic life is based on a complex *information code*, and such a code cannot be created without a Master Designer at the *cosmic keyboard*.

CHIRALITY

Life to have evolved requires intelligence. No amount of certain molecules in the right places, no amount of high voltages latent in natural lightning strikes impinging upon a primordial soup, can ever result in highly complex, organic life forms. A mixture of molecules using clever chemistry to link them up in just the right order does not in itself produce a molecule latent with life's principles.

Nature is symmetrical, at least at the molecular level. The word "chiral" comes from the Greek, *cheir,* which means *hand.* Practically everything can be identified as either left-handed or right-handed in our world today. Chiral molecules are mirror images of each other as illustrated in Figure 4.4[4]. Left-

handed *amino acids* and right-handed *sugars* are chiral, necessary for all living systems. They *must* be in pure left- or pure
right-hand forms, never in a mixture of both, else they would
be lethal. There are no exceptions!

Here is the conundrum: If normal chemical reactions *cannot* produce the required pure forms, pure right- and pure left-
hand forms, where did the very first DNA molecule in existence receive its information in order to arrange itself into the
proper pure right- and pure left-hand forms?

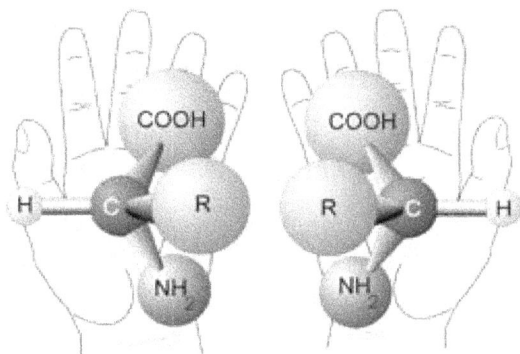

Figure 4.4: Chirality

Information latent in the DNA molecule must have originated in God who only possesses the ability to communicate
specific instructions to life's beginnings. The chiral acids and
sugars in the molecule give the DNA helix its characteristic
right hand spiral twist, leading to an observable 3-dimensional
shape. The twist is crucial to life and only one component containing an opposite chirality would render the whole lifeless.

CHAPTER THOUGHTS

The DNA molecule invades everything that requires life. If
you have a need to create life, then you must start with the
creation of the DNA molecule (I strongly suggest you not try it).
But let's try it anyway! If we can do it, then anyone can.

This is not going to be a simple task. You might start with
carbon, the chemical element crucial to all life as it exists in
the world today. Then you might try to add hydrogen and oxygen to the mix. Put it all in a pot, turn on the mixer, and hope
for a miracle that combines the elements in the correct order.

The first thing that must happen is that five carbons must attract and mate with ten hydrogen and four oxygen to somehow come out with what we call a pentose sugar molecule; and after that we could add one phosphorus and four more oxygen. Now don't forget, you must remember that the sugar must be all right-handed. One left-handed sugar would spell disaster. Start over again and do it correctly this time!

While that's cooking, let's try an A amino (adenine). Take five carbons, five hydrogens, and five nitrogens and hook them up in their correct positions, remembering that all aminos must be left-handed—or else! Once you obtain the proper configurations, mate A (adenine) on rail 1 with T (thymine) on rail 2—likewise, continuing to the bottom, mate G (guanine) with C (cytosine)—and hold them together with hydrogen bonding. How? Why ask me! Look at Figure 4.2. You figure it out.

Nothing is working here. It looks like there is something missing, some sort of *external source of energy*. We don't seem to have the right mix. Nothing is bonding together as it shows in the text books. Maybe the sun is not strong enough today! Or maybe we need the same kind of energy that the producers of the old-time movies used when they created guys like Frankenstein—you know, like high-powered lightning strikes.

The recipe that we are using for the DNA molecule seems to be incomplete. Seems like *information* is missing, and how do we go about putting all this soup together, this DNA broth? What type mixer can we use? Is this the only *conversion process* available? It seems like everyone working on this problem of life is having the same kind of difficulties. When we do get it to work, how do we get the genes that we want out of the mix that contains 35,000 of them? What are the specific genes that make the proteins that will make us a new kidney, or a new heart, that we want to use in a medical transplant?

I give up—this whole project is too complicated. The trash heaps of the world are full of spoiled recipes, authored by all those who are uninformed and remain in unbelief, who consistently try to tell us of "How to Bake a 3-Layer DNA Shortcake in Three Easy Steps."

5

From DNA to People

"Wisdom is the principle thing; therefore get wisdom: and with all thy getting get understanding" (Proverbs 4:7)

To understand life, one must understand the DNA molecule. To understand the DNA molecule, one must seek the *wisdom* of God, the *principle thing*, and with wisdom *understanding* is sure to follow. King Solomon never penned a more accurate and truer verse than that quoted. Take heed then, less we succumb to the fallacy and wiles of human wisdom lurking behind every event. Therefore, we now leave the elementary principles, however fascinating, and forge onward to grasp some of the functionality of this amazing molecule.

Transcription

Genes code for proteins and in protein production amazing machinery is put into place. To synthesize or to make a protein from a strand of DNA, a cell must be able to access the *information* contained in the blueprint impressed upon the gene. *Gene Expression* is the term scientists use to describe the two-step process that occurs at the molecular level. Ribonucleic Acid, RNA, is the molecule that is responsible for the operations within the nucleus of the cell. The structure of the RNA molecule differs from the DNA molecule in three observable ways: 1) DNA molecules consist of two parallel rails with connecting base pairs, while RNA molecules consists of only one rail with unpaired bases; 2) RNA substitutes the base U (uracil) for the base T (thymine) as in DNA; and 3) RNA contains one extra oxygen atom, making all the difference between life and death. This *open-ended* feature of the RNA molecule has the propensity to react with molecules by base pairing with its complimentary base on the DNA strand.

FROM DNA TO mRNA

DNA and RNA are seen as Siamese twins that are inseparably joined together at the hip, as was Naked Truth and Bare Facts in our discussion of caterpillars and butterflies, for wherever DNA goes, RNA is sure to follow. DNA stores the instructions for making things like you and me, but before your cells can use DNA it has to be "transcribed" into something they can use. To do this, the DNA strands pull apart (unzips) and RNA, which is a single strand, comes in and matches up with the bases on one of the open DNA strands.

A *messenger* RNA (mRNA) molecule is constructed using a gene sequence as a template and copied into the nucleotide sequence of the mRNA molecule. Only one strand of DNA is transcribed; this is the "coding strand."

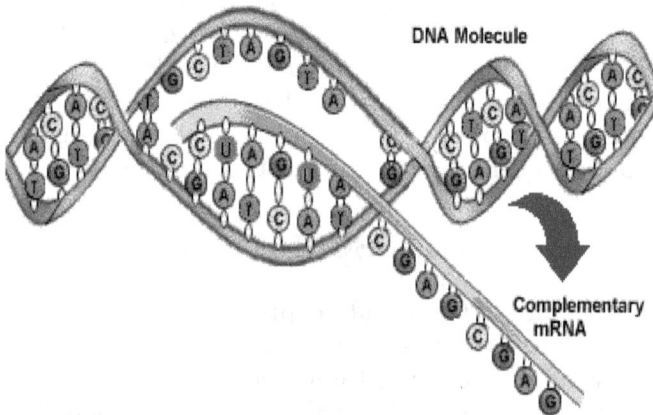

Figure 5.1: Transcription of mRNA

In Figure 5.1[1], the *helicase* enzyme (the gate keeper) unravels the helix, unzips a section of the double stranded DNA molecule containing the desired gene, breaking the weak hydrogen bonds between the complementary base pairs in the unzipped section. Other specially designed enzymes must immediately keep the two single strands apart while they are separated, else, they would immediately zip back up again.

Other enzymes then match a base on the DNA molecule to a corresponding, complimentary base on the mRNA molecule—and keeps on adding and elongating the mRNA strand by adding *free-floating* bases to the complimentary strand.

Hence, the coding strand is *transcribed*, using the sense strand as a template. As shown, the DNA coding strand in the unzipped section reads GAT CAT. This DNA sequence would normally bond with CTA GTA, but when considering mRNA, A on DNA bonds with U (uracil) on mRNA and not with T as on another DNA strand. Therefore, the complementary mRNA sequence will read CUA GUA, not CTA GTA. Finally, after the complementary mRNA is made, the open section of DNA is zipped back up into its original helix formation. Simple, isn't it!

TRANSLATION

In order for the desired protein to occur, the gene sequence of DNA nucleotide bases transcribed onto the mRNA molecule must be *translated* into the amino acid sequence of the new protein, using the code impressed upon mRNA.

FROM mRNA TO PROTEINS

Unlike DNA, mRNA molecules are free to float out of the nucleus of the cell via pores in the nuclear membrane into the cell body, for it is in the cytoplasm of the cell where translation takes place. At this point mRNA, ribosomal rRNA, and transfer tRNA, all come together.

mRNA carries the DNA nucleotides, Figure 5.2[2], and attaches to ribosomal rRNA, the "factory floor" of the molecular assembly line. As mRNA travels to the left, Methionine with bases AUG, is the start codon that tells the system to *start* reading code here; then Arginine (CGU), and Tyrosine (UAU), with Threonine (ACG) yet to be read by rRNA on mRNA.

The code always ends in one of the stop codons, UAA, UAG or UGA, that signals the process to *stop* making the protein. There are no punctuation marks in the code and it is read continuously every three bases, forming amino acids as it goes along, and stops when the desired protein is produced.

There must be three bases in a codon; one or two bases will not specify a protein—not enough information is available. Four is sufficient but not necessary. There are 20 different amino acids used in our body and the order determines the specificity of the protein. One out of order is never beneficial.

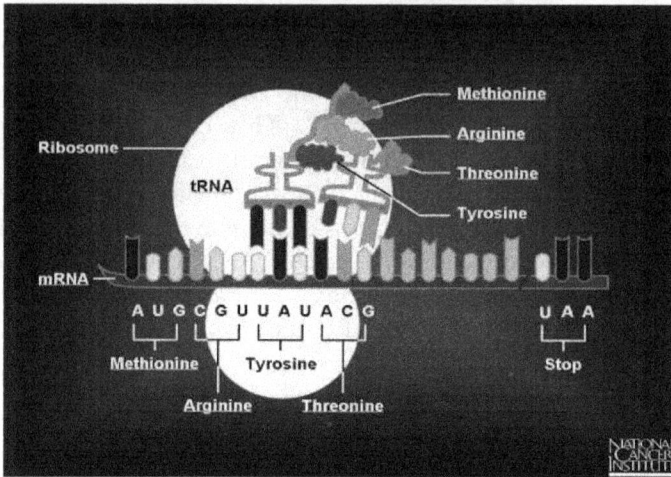

Figure 5.2: Translation of mRNA

As the mRNA travels to the left, rRNA first reads the start codon AUG, singularly (A)denine-(U)racil-(G)uanine. The next reading of CGU initiates a response from tRNA, the factory worker that pick up free-floating anticodons in the cytoplasm, asking for its amino acid code. This procedure continues in an assembly line fashion until rRNA comes upon one of the stop codons. By reading the entire mRNA sequence, rRNA and tRNA construct a long chain of amino acids that make up the required protein.

In a nutshell, translation involves taking the message in DNA, transferring it to messenger mRNA, decoding the message from the language of amino acid nucleotide bases to the language of amino acids in proteins. It is seen that once dissociated from its DNA sugar and phosphorus groups, it's the amino acid codes that carry the messages of life forward.

Confusing? You bet it is, but try to hang-in there!

THE CELLULAR LEVEL

Theories and observations of genetics span four levels of biological organizations. We have discussed in detail the first of the levels, the Molecular, where cancer and predestination genes reside, and where gene expression, consisting of transcription and translation, occurs, resulting in specific proteins that are highly functional. We have seen that we, as human beings, are not really flesh and bones at this level, but are of atoms, bonds, sugars, and bases, and things called nucleotides.

In the Cellular level, amazing microscopic compounds exist and have their being in a world all its own. If we want to know about life itself, of how the body works, we will need to know about the human cell and how the cell works. It seems that everything that happens to us happens to us at the cellular level. If it doesn't happen there, it will never happen!

THE CELL

But now, look at yourself in a mirror—what you see?—not atoms but about 60 trillion cells divided into about 200 different types, performing 10 million reactions per second. Muscles are made of muscle cells, livers of liver cells, and kidneys of kidney cells. There are very specialized types of cells that make the enamel for our teeth, the clear lenses in our eyes, and those dirty nails hanging on the end of our fingers and toes. All are different in appearance but similar in composition and function.

The human cell, figuratively in Figure 5.3[3], is a stand-alone living entity able to eat, grow, and faithfully reproduce its exact kind. The cells in your body are composed of complicated microscopic machines, each having a specific function to perform. The brains behind the cells are the DNA chromosomal molecules tucked away in the nucleus.

Your genome is the book, the genes are the words that make up the sentences, and the letters are the four nucleotide bases. Think of it; four letters in various sequences is all that is necessary to construct something as wonderful and marvelous as a human being. All the necessary information is there.

Your DNA is like your thumbprint—it is yours and yours alone. Unless you have an identical twin, no one else on the planet has exactly the same DNA as you. In the nucleus of every cell in your body is the collection of DNA needed to make you. DNA in the nucleus is grouped into 23 sets of chromosomes, 46 individual DNA molecules called the *genome*. In each genome the DNA is grouped into genes, about 35,000 of them, each carrying information that tells the cell to make a unique protein that can perform special functions.

Cells carry sufficient information to construct a living person, of where and how to place each organ with differing bodily functions, of what gender you are to be, to make blue eyes or brown eyes, curly hair or straight hair, or no hair.

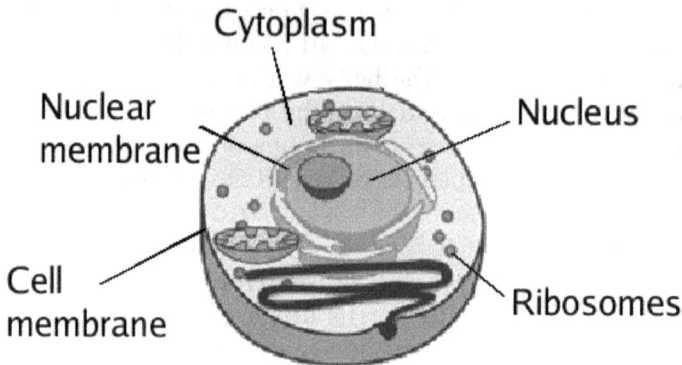

Figure 5.3: The Cell

Down through the annals of Biblical history, men are seen always to be *begetting* sons and daughters—*Abraham begat Isaac; and Isaac begat Jacob; and Jacob begat Judas and his brethren* (Matthew 1:2)—all the way down to our present-day family lineage. Drilling down deeper, God, in His wisdom, wisely neglected to tell us all the facts contained in the creation narratives of Genesis. Allow us then, to expand the narration to include:

In the beginning God created *Atom*: And Atom begat Carbon Bonds; and Carbon Bonds begat Bases; and Bases begat Nucleotides; and Nucleotides begat Aminos; and Aminos begat Genes; and Genes begat DNA; and DNA begat Chromosomes; and Chromosomes begat mRNA; and mRNA begat rRNA; and rRNA begat tRNA; and tRNA begat Proteins; and Proteins begat Cells; and Cells begat Organs; and Organs begat *Adam*; and Adam begat....*You and Me*!

JUNK DNA (jDNA)

Not all the nucleotide bases in a gene are used to express a particular protein. Genes can be divided into sections that code and sections that do not code for proteins. Those sections that code and are useful are called *exons* and those that do not code are called *introns*.

Some scientists regard introns as Junk DNA (jDNA) and have been used as an argument against intelligent design and for the random process of evolution (God forbid!). Others suggest they may have some as yet undiscovered function. Whatever their function is in the genome, the Creator, whose intelligence far surpasses that of humans, would not clutter up the genome with 97% nucleotides that have no function at all. On the other side, how could only 3% of the genome code for life with all its amazing functionality and flexibility, and have the rest of the genome do nothing! God is a God of purpose and usefulness, not a God of randomness and uselessness!

It is our position that jDNA plays a vital function in the creation of the embryo of a human being, and is only the tip of the iceberg of molecular biology. jDNA is too uniform and patterned to be of no value. This suggests that they have some role to play in regulating the development in the life and character of a Christian. God has gifted those in His created *kinds* with the ability to diversify, to adapt, and that under a high degree of organized control.

Portions of the genome may remain stable while some parts of the same chromosome may be widely subject to variations. The end result of all this is that new and previously unused proteins may come into play; building blocks of a new human design—a new species—a *born again Christian!* In so doing, new information to guide the process needs to be added or existing codes altered, thereby issuing-in a new life. "Junk" DNA is some rather amazing "junk."

CELL DIVISION

The lifetime of a human cell is given in terms of how many times the cell will divide and thus reproduce itself. The blueprints for cell division are contained in the chromosomes and the initiation of cell division is itself the division of the DNA genome. Each time a cell divides, producing two daughter cells, the entire DNA in the genome must be faithfully *replicated*.

Replication is the faithful copying of the DNA genome, producing two new strands of DNA. The helicase, the gatekeeper, unravels and then unzips the two DNA strands producing a *replication bubble* as it moves in both directions up and down the molecule.

The two old strands serve as a template for the making of the two new strands. In this manner, each daughter DNA molecule is an exact replica of its parent DNA molecule and each contains one strand of the original parent.

Here is the conundrum: since cell DNA replicates and furnishes all the necessary information for the cell to divide, and divide, and divide, why should we die? Why not live forever? After all, all single cell organisms, such as amoeba or bacteria, live-on endlessly and effortlessly. However, the human body, like all multi–cell organisms, grows up and then starts growing down, eventually to die. Why do we have to deteriorate as we grow old? Bacterial forms do not have to suffer that fate so why, you might ask, must we!

CHAPTER THOUGHTS

A DNA molecule, with the assistance of midwife Helicase, gives birth to a son, mRNA, and christens him Messenger. Messenger then gives birth to two siblings: rRNA—Ribosomal, and tRNA—Transfer, grandsons of DNA. The RNA family members are constantly at work performing acts of gene expression, transcription, and translation, never resting from their duties, always producing offspring.

It all starts out with replication, the basis for why we look like our parents, a process occurring in all living organisms, to copy their DNA. This process involves each strand of the DNA molecule serving as a template for the reproduction of a complementary strand. The result is that an identical DNA molecule has been produced.

Cells in the human body exist for a very short time and must be reproduced continually, making proteins from many genes. Although genes get a lot of attention, it is the proteins that perform most of life's functions. Unlike the relatively unchanging genome, the dynamic proteome changes from minute to minute in response to tens of thousands of intra- and extracellular, environmental signals.

Implications in research are hopeful and reassuring. It is believed that cells may have the biological potential to carry us into our mid 100s in terms of longevity. But is this all that we can hope for?—to live a mere 150 years and then, what? Die? There must be something better in this life, something that affords the possibility of unlimited extension of life (dare we say—*immortality?*).

6

BIOLOGICAL IMMORTALITY

"For this corruptible must put on incorruption, and this mortal must put on immortality" (1 Corinthians 15:53)

One of the hallmarks of a cancer cell is its immortality—its ability to live and divide endlessly. This concept is as easy to understand as it is hard to implement. We speak here about the infinite continuation of human life in the same form as we know it, based on the same biochemical processes in our bodies that make us alive today. There is a mechanism of aging and death that is built into our bodies by nature and is referred to as "programmed death." If we could somehow switch off this mechanism, in principle, we could live indefinitely. The body has a capacity of self-renewal, a process of life that could be unlimited in time.

TELOMERE: THE BURNING FUSE

In the biological world, there is an alluring temptation to associate immortality with the structure of the DNA molecule. At the tip of each end of chromosomes are very short sections called *telomeres*. If you do not know what telomeres are, you are in for a big shock. The length of one's life given in years depends upon, at the cellular level, the length of DNA occupied by telomeres. Normal telomeres are tandem repeat sections of DNA sequences consisting of six bases of TTAGGG on one strand; and base paired with its complement of AATCCC on the other strand; occurring about 12,000 times, or 2,000 repeats, at each end of all chromosomes. Remember: T with A, and G with C. A cell needs to divide in order to preserve the organism that it supports. In normal cells, the rate of new cell growth is kept in balance with the rate at which old cells die. The problem begins when normal cells divide. Helicase (the gatekeeper) unzips the DNA molecule and an enzyme then moves in to pair up the two strands with its DNA base pair complement. This procedure, however, does not supply all of the added bases needed to complete chromosomal division. Some mating bases are *missing*.

> A telomere is a region of highly repetitive DNA at each end of a linear chromosome that functions as a disposable buffer. Each time a cell divides and each chromosome replicates, or makes a new identical copy of itself, the process can not extend to the ends of the chromosome thereby leaving a section at the ends with no mating bases. Instead of ending up with a chromosome that is entirely composed of nucleotide base pairs, the ends are now single strand DNA with non-mating bases and not the usual double strand.[1]

Telomeres, crucial to the life of the cell and the key to long life, eternal life in our quest, sought after by men of all ages, lie in the repeated base sequence of TTAGGG *faithfully occurring*. However, each time a cell divides, it is estimated that its telomeric DNA shortens by about 100 base pairs, representing 16 TTAGGG repeats. Not only do we lose DNA by this action but we also loose *information* necessary for the maintenance of life. Little by little, as the cell divides and increasingly becomes abnormal, diseases and old age are steadily creeping in. When the cell divides by about 50 times, the Hayflick Limit is reached, and the cell will start to die; and after 120 divisions, telomeric DNA will completely vanish and apoptosis will set in—programmed cell death, hence "the burning fuse."

The cell is programmed to stop dividing before it reaches the 50 divide limit. At that point, there would not be a continuation of new cell divisions to reach the limit, thereby overcoming the effect of cells lost by dividing and replenishing, resulting in old age and beyond. The key to longevity is to keep the telomeres *long* and not allow them to shorten after each cell division. If telomeres could be kept long, not loosing cells, we would be able to continually reproduce new, sustained cells and thereby live forever, hence the quest for immortality.

TELOMERASE

An enzyme that can repair and lengthen the effects of telomere shortening is called *telomerase*, used as a template to guide the insertion of nucleotide bases to the overhanging repeat sections of TTAGGG on the lagging strands. By lengthening this strand, *polymerase* is able to complete the synthesis of the incomplete ends thereby maintaining the original length.

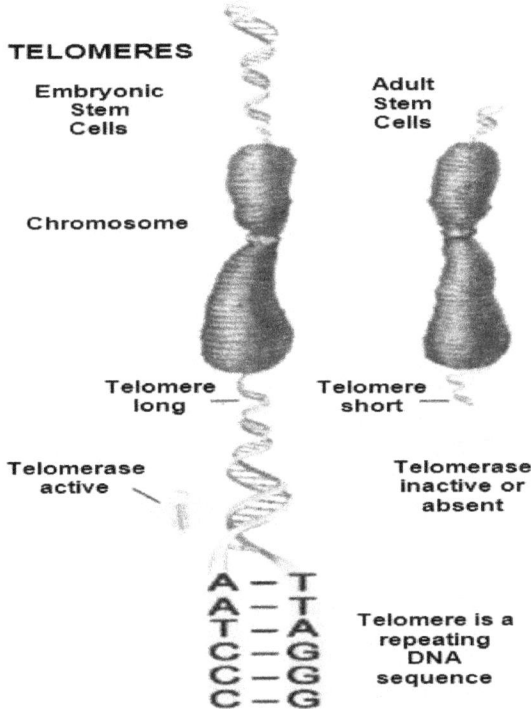

TELOMERES

Embryonic Stem Cells

Adult Stem Cells

Chromosome

Telomere long

Telomere short

Telomerase active

Telomerase inactive or absent

A — T
A — T
T — A
C — G
C — G
C — G

Telomere is a repeating DNA sequence

Figure 6.1: Telomere Shortening

Figure 6.1[2] graphically illustrates the major players in this process. The upper portion of the diagram illustrates both embryonic (germ line) and adult (soma) chromosomes with their individual telomere sections protruding from both ends of the chromosomes. The length of the embryonic telomeric section is *long and normal* compared to that of the adult section, which is *short and abnormal*. This is always the case because the repair enzyme telomerase is active in the embryonic mode and inactive in the adult mode, keeping the original length of the telomeric section constant. But, one problem immediately arises; cancer cells thrive on telomerase!

That is one reason why cancer cells proliferate without bound. So, here is another conundrum: do we perform telomeric therapy to increase longevity at the expense of feeding cancer cells and dying of cancer, or do we limit telomeric therapy to decrease the rate of cancer cells but increase the rate of physical mortality. If we were to produce enough telomerase to keep our cells living, we would still have to combat the myriads of other health problems that would kill the immortal cells anyway. Defeat cancer, and other strains of viruses might show up that we would need another cure for. Beat all that and you will still have the occasional driver who runs you into the ground; seclude yourself from everything and you will die of boredom. So, what good is this type of immortality?

HeLa

Her name was Henrietta Lacks and she died in 1951 of cervical cancer at the tender age of thirty-one. Henrietta was married, had five children, and by all indications lived a full and happy life. Members of her family all praise her life as dedicated to God and lived as if she really meant it. Henrietta was laid to rest in an obscure section of a rural cemetery, in an unmarked grave across the street from her family's tobacco field, behind the house where her mother was born. Her family and close friends remember her well, but little is known or remembered of her today by neighbors in the community along the one-block, unmarked road called Lacks Town Road, in Clover, Virginia. But there is where the similarities end as with others of her time and of the same way of life, for although Henrietta Lacks died over a half century ago, Henrietta is *alive and well* today, and *still producing offspring!*

As the story was told, and printed many times in the newspapers and magazines[3] in many different ways, during her surgery for removal of her cervical cancer, a doctor removed a small section of the tumor for chemical and biological analysis, as all hospitals do under normal procedure. Eight months later, Henrietta Lacks died, and it was not until more than twenty-five years later that the subject of what happened to, and what were the results of, the medical analyses of Henrietta's biopsy sample.

Now by this time it was going around that something called HeLa cells were finding their way around town, all the way around the entire world for that matter. Doctors and scientists in hospitals and research centers clamored for HeLa cells that have been cultured *in vitro*, in test tubes—from the U.S. to Europe, to Russia, to Asia, and then back again. Colleagues used HeLa cells to grow the poliovirus that was ravaging children throughout the world, and even found their way into space that multiplied in a shuttle far above the earth, as scientific experiments were conducted to determine how the cells would behave under no-gravity conditions.

Scientists in every quarter of the globe today are working with, you guessed it, Henrietta Lacks' HeLa cervical cancer cells that were obtained from the biopsy taken from her cervical cancer. These particular cells, named HeLa for the first two letters of her first and last names, proved spectacularly successful. They *do not die of old age* and can divide an unlimited number of times as long as basic cell survival conditions are met. There are many strains of HeLa cells, but all HeLa cells are derived from the same tumor cells removed from Henrietta Lacks.

The day research scientists got their hands on Henrietta Lacks' cervical cancer cells, everything changed in the medical world that eventually would spill over into the theological arena. Henrietta's cervical cells multiplied like nothing that anyone had ever seen before. They were the fastest growing cancer cells ever known, and they reproduced an entire generation every twenty-four hours. There is more of Henrietta now, in terms of biomass, than there was when she was alive! Can we now say that HeLa cells are "biologically immortal"?

HELACYTON GARTLERI

Henrietta's cells were the first human culture to survive beyond the 50th generation, and, in fact, some biologists believe that HeLa cells are no longer human at all and consider them single-celled micro-organisms. They express telomerase during each division, thereby circumventing the Hayflick limit, preventing aging and eventual death. This suggestion however has not been followed by many biologists in the belief that a chimeric human cell line is not a distinct species. Or is it?

In 1991 the scientific community decided it was, and blessed HeLa cells with its own genus and species: *Helacyton gartleri*. That would make Helacyton gartleri an example of speciation, when a new species is observed developing from an existent species (may we say: such as a *butterfly coming from the caterpillar*, or as a *Christian coming from an unbeliever!*).

SPECIATION

Many of us might like the idea of living forever instead of having our bodies decay away after only a few years. The biological ticking bomb in our bodies is the caps of our chromosomes, keeping the ends of our DNA from unraveling and unzipping—God's way of making sure that a species does not grow out of control.

Viruses, free radicals, and even stray gamma rays break apart DNA all the time, and sometimes the result is mutation. If the mutation survives and lives as well as the surrounding cells, we call it cancer, and it usually kills its host, and when the host dies, so do the cancerous cells die—usually!

But not so with Helacyton gartleri. These cells will live forever in culture and will never die. Their telomeric sections are constantly kept at the proper length by the action of telomerase. It does not take a rocket scientist to discover the benefits of participating in the formation of a new, *immortal* species, particularly a species that is not able to die, regardless of all circumstances imposed upon it.

However, is such a species obtainable? What conditions must be met in order to obtain such a status as to live forever? Between life and death we know a great deal, but outside these limits, in both directions, we know exceedingly little. We know that because of the sentence of death imposed upon all of humanity, we are all programmed to die as evidenced by all who reside in the cemetery. The prophet Job echoed the hope of gaining immortality centuries ago: *If a man die, shall he live again? all the days of my appointed time will I wait, till my change come* (Job 14:14), and then supplies the only viable answer to our dilemma:

> *And though after my skin worms destroy this body, yet in my flesh shall I see God, Whom I shall see for myself, and mine eyes shall behold, and not another; though my reins be consumed within me* (Job 19:26–27).

Is there any possibility of immortality *now* or must we wait for death to achieve this new state of existence, where we will not be able to die no matter what happens to our body. It is not clear if it is possible to modify the body in such a way that we can become immortal and live forever. Can we switch off the mechanisms that are at the root of this shortening action, causing aging in all of us and eventual death? Is it already too late? Current theories in biology do not give us a definitive answer to this perplexing question, nor does it offer any assurances that an answer will soon be forthcoming.

Both the Bible and the Second Law are against this position as nothing in this universe is immortal, as death touches all living creatures. The Bible is emphatic in stating that man will eventually die because of the penalty of sin therefore, that settles it. However, for the Christian in this life, now that is a different matter altogether, for in order to become immortal we must go beyond our present form of life.

Just what would that present—*form of life*—be?

CHAPTER THOUGHTS

Mortality in this life is telomere shortening, a few bases on the end of our chromosomes that can't seem to get their act together. They short-change the chromosome by leaving one DNA strand short of supplying the required number of bases for a long and happy life. Telomerase seems to want to correct this abnormality, rectifying the situation by furnishing the required free-floating bases to close the gap.

But wouldn't you know it—cancer cells love to munch on the tasty telomerase! This leaves us with the conundrum of dying with cancer or dying by apoptosis—programmed death. Which do you prefer?

Henrietta Lacks, bless her heart, wouldn't have any of this. HeLa cancer cells are alive and well today, producing offspring all way around the world. Just keep feeding her and she will keep on doing what she always does—live! But, is Henrietta *actually* alive today, sitting in test tubes all around the world?

Come-on now: Is Henrietta alive or is she dead? You make the call.

7

DESIGN! DESIGN! DESIGN!

*"I will put my laws into their mind, and write them
in their hearts"* (Hebrews 8:10)

Allow me at this point to introduce to you the most fantastic design that our awesome God has incorporated into the biology of the human body, governing the way we think and act to the minutest degree. We refer here to *neuropeptides and receptors*, and attempt to correlate both entities to regeneration, emotions, immunopeptides, and to viruses. Connect both neuropeptides and emotions together and out comes what scientists today refer to as *molecules of emotion.*

Many people perceive that there is a great gulf between science and religion, that these entities do not intersect and that one is distinct from the other. Christians, however, should recognize and assert that science and religion are compatible, that the physical and the spiritual are, indeed, interrelated.

This can be seen in our everyday activities. Human behavior is genetically determined and played-out in the human body by means of biological actions and reactions. That there is a strong genetico-spiritual component in our behavior is becoming increasingly evident. *Environmental forces external to the human body exert powerful effects by modifying genetic structures and hence, body chemistry.*

Emotions are the key to life, and a good place to start to define what we mean about "molecules of emotion" is to start with the nervous system. Our body is loaded with nerves. But what is the function of nerves and how do nerves work (don't worry, for they also consist of atoms, molecules, and DNA!).

THE NERVOUS SYSTEM

You touch a hot object; you yell "ouch," and immediately you drop the object and pull your hand away from the heat source. You do this so quickly you don't even think about it. How does this happen? Well, your nervous system coordinated everything; all the thoughts, movements, and sensations that you experienced. It sensed the hot object and then immediately signaled your muscles to let it go. This system consists of two main branches, the Central Nervous System (CNS), and the Peripheral Nervous System (PNS).

THE CNS

The Central Nervous System is essentially a biological information highway, and is responsible for controlling all the processes and movements in the body. It can receive and store information, interpret results, and execute final responses. The CNS unconsciously performs the processes that are executed in the brain.

The CNS is the center of the nervous system consisting of the brain and spinal cord. It is responsible for receiving and interpreting signals *from* the PNS and sends out signals *to* the PNS, either consciously or unconsciously. Due to the importance of the CNS, it is encased within the cranium protecting the brain, and within the spine protecting the spinal cord. The CNS is arguably the most important part of the body as it controls biological processes and all conscious thought.

THE PNS

The Peripheral Nervous System branches off into the Somatic Nervous System, (SNS), that portion concerned with the control of voluntary muscles, i.e., movements that you consciously thought about doing; and the Autonomic Nervous System, (ANS), incorporating all the impulses that are done involuntarily, and are associated with essential functions such as breathing, heartbeat, etc.

Of primary importance are the Sympathetic and Parasympathetic divisions that branch off the ANS system, Figure 7.1[1], that keep one another in check in a form of negative feedback;

that if one system over-reacts to a stimulus, the other system compensates to return to a mutually agreed upon level.

Let's now look at the structure and functions of the CNS and ANS and how nerve cells communicate with each other and with various cells and organs within the human body.

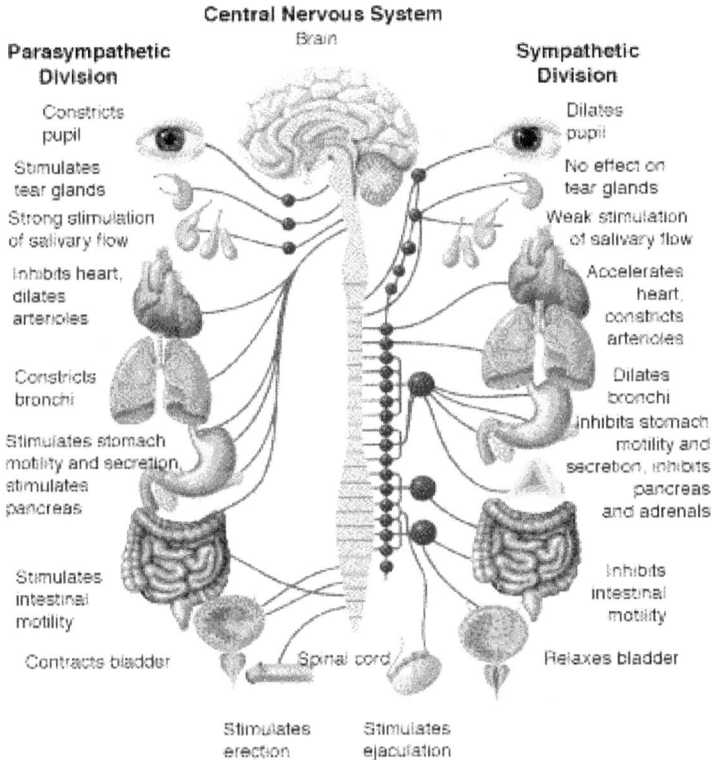

Figure 7.1: The Nervous System

THE NERVE CELL

The brain is the center of the CNS like the microprocessor in a computer. Nerves are likened to the computer wiring connections. Nerves carry signals to and from different areas of the nervous system as well as between the nervous system and other tissues and organs. The brain, in conjunction with the spinal cord and nerves, are the body's informational highway that does all the work of processing and communications.

Neurons, Dendrites, and Axons are what the brain is all about. The adult brain, weighing only three pounds and averaging 1400 cubic centimeters, contains about 100 billion neurons. The neuron or nerve cell is the basic building block of the brain. In the formation of a human body, neuron cells form at an astounding rate of 25,000 per minute! Each neuron cell contains an extremely complex network of branching fibers, called dendrites, and each neuron cell is said to be in *dendritic connection* with as many as 10 thousand other neuron cells. The total number of dendritic connections, also called "bits," is of the order of 100 trillion, and if they were laid end to end, they would circle the earth more than four times, or over 100 thousand miles! An analogy by M. Denton is illustrative:

> Imagine an area about half the size of the USA (one million square miles) covered in a forest of trees containing ten thousand trees per square mile. If each tree contained one hundred thousand leaves, the total number of leaves in the forest would be 100 trillion, equivalent to the number of connections in the human brain.[2]

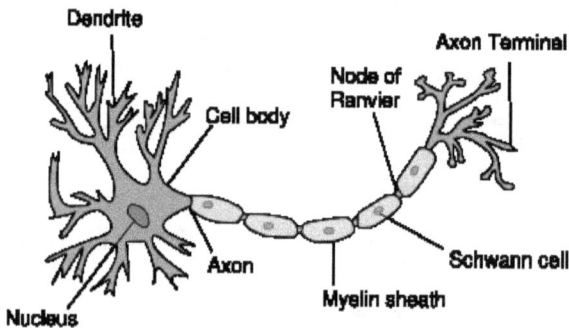

Figure 7.2: The Nerve Cell

Each neuron has one axon, Figure 7.2[3], a long, thin cable-like projection that transmits nerve *messages* along the length of the cell. The cell has many dendrites, nerve endings, small branchlike projections that make connections to axons of other cells and allow neurons to receive messages and thus are able to *talk* with other cells.

Neurons gather and transmit signals to and from the brain and nervous system at an astounding velocity of up to 300 miles per hour. They have the same characteristics and parts as other cells, such as DNA and nucleus. They transmit signals and pass messages over the entire length of the body, from neuron to neuron, from neuron to muscles, from neuron to glands—back and forth and then back again.

Axons are responsible for transmitting outgoing cell messages and dendrites do all the receiving. Neurons *never* directly connect with each other. If neuron A sends a message to neuron B, it must transmit along axon A. But, neuron B cannot receive the message on axon B, therefore a dendrite B must be used to receive the message from axon A. But, between axon A and dendrite B, there is a very small disconnect, a gap, called the *synapse*.

Everything happens in the synapses, the small disconnect between axons and dendrites. Now, if there is a gap disconnect there, from axon A to dendrite B, how does the *information* contained in the message *jump* across the synapse? Not only that, but just what is this message? What is it that does the jumping? What does it say? How does it talk? This is all weird!

Neurons are everywhere. Since there are 100 billion neurons, then there are 100 billion axons. Don't ask how many dendrites there are! And please, don't ask me how many receptor cells are on the surface of each dendrite! I believe that if you were to count all the stars in the heavens, you wouldn't even come close.

THE BRAIN

Whether considering a star or a starfish, an infinitesimal virus or an infinite universe, the concept of design remains one of the strongest evidences in support of, yea it cries out for, an Intelligent Designer, fulfilled only in the reality of a Supernatural Deity. There is one particular aspect of design that is so powerful, so convincing, that it almost seems unfair to challenge unbelievers with it. The reference is to the human brain, the greatest concentration of chemo-neurological order and complexity in the physical universe. It is a video camera with the storage capacity of libraries, a computer of immense power and speed, and a communications center capable of handling vast amounts of information, all in one 3-pound organ! And the more you use it, the better it becomes!

The *cerebrum* is the largest part of the brain. The surface—the *cerebral cortex*—rich in dendrites, lies in the *gray matter* that sit on top of a large collection of *white matter*, rich in axon pathways. The entire cerebrum is divided into two halves, cerebral *hemispheres*. Located deep within the cerebrum are the *thalamus*, the *hypothalamus*, and the posterior lobe of the *pituitary gland*. The spinal cord is an extension of the brain, and together with the brain, forms the central nervous system. It originates at the bottom of the brainstem, runs down the central canal of the spinal column, and extends two-thirds of the way down the spine.

CELLULAR COMMUNICATIONS

Enter now into the strange world of neuropeptides and receptors! Listen closely! Can you hear your cells chatting with each other? Some may be talking about the football game last night that you watched on TV, or others may be reminiscing about the cute girl that moved in next door that you took a fancy to. Whatever they are talking about, you can be sure of one thing—they are talking! And they never cease to talk. They talk day and night, rain or shine, awake or sleeping—you just can't keep them from talking. Think on this; if your neuropeptides and receptors are not talking about her at the molecular level, you can never talk about her at the populational level!

NEUROPEPTIDES

Everyone is familiar with writing messages in a word processor on a computer without ever thinking of how the computer accomplishes this task. In radio and television, messages are both in analog or digital format, encoded onto a carrier frequency and then transmitted to other locations, where a receiver tuned to that frequency picks up the signal; decodes it and lo-and-behold out comes the audio music and video picture.

So also in biology, there is no difference. Countless messages are received throughout the human body and retransmitted to other locations within the body without anyone ever thinking twice about how it all works. Neuropeptides and receptors are at the root of this amazing message transmission, reception, and transfer system. The main difference is that they are *alive*—I think! Life is cells talking to each other.

Life without cellular communications is cellular death. Cells do just about everything. They detect, interpret, and respond to chemical signals, just as we respond to audio signals coming from the radio, from conversations over the telephone, or from a one-on-one conversation with friends.

Cells are able to talk because of neuropeptides that contain the chemical *language* of cells, able to communicate across body systems, and to regulate all life processes on the cellular level. They extend to every corner of the body and function as a living processor of information, a means to transmit messages across organs, tissues, cells, and DNA. They control our mood, energy levels, pain, pleasure reception, body weight, and our ability to solve problems; they also form memories and regulate our immune system. WOW! What a system!

Let's stop for a moment and think about talking. When you say "hello" to someone, who is doing the talking? Normally you would reply—"I am." Really? Is it your mouth, your tongue, your throat, or maybe your vocal chords that are doing the talking? Going down into the cellular level, it looks to me that it is your neuropeptides talking to specific receptors located in your body, transmitting the cellular message of "hello."

Neuropeptides are the messengers of the brain, mobile signaling molecules, short chains of amino acids (we've seen these little critters before). Aminos containing only one acid is classified as a neurotransmitter, from 2-50 are neuropeptides, from 51-200 are polypeptides, and above 200 are classified as proteins. Some proteins number in the thousands.

DNA in the nucleus of the cell determines which amino acids are to make what neuropeptides, in processes identical to mRNA transcription and translation in making standard proteins. Neuropeptide messages are transported from the cell body, through the axon to the terminal of the cell to await further action. When a cell is stimulated to release the peptides, the contents are emptied into the synaptic cleft, the space between the axon on the transmitting side and the dendrite on the receiving side.

After the neuropeptides have been released from the cell terminal, different things may happen. Either the molecules are broken down instantly, transported back into the terminal, or they go on to exert an action elsewhere. They are most likely to move through extra-cellular space, carried along in the blood, traveling long distances, causing complex changes in the structure of remote body cells.

Neuropeptides are informational substances that coordinate almost all physiological and emotional processes on a cellular level....Therefore, it is suggested that neuropep-

tides circulating in the CSF may be the physical substances that coordinate the mind, body and Spirit connection and facilitate healing on all levels.[4]

Neuropeptides are the molecular basis of emotions that
control every system in our body. A neuropeptide that carries
the message binds to a specific receptor located on the surface
of a cell, Figure 7.3[5]. This action produces secondary messengers inside the cytoplasm of the cell that ultimately results in

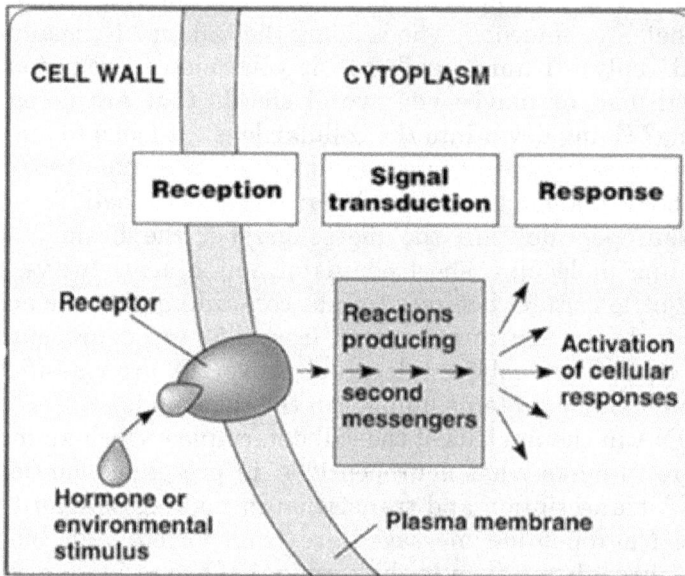

Figure 7.3: Cell Communications

cellular responses, including emotional and physical changes.
In all cases, a cell surface receptor must be present in order to
receive and bind to the peptide. Single or multiple cell receptors determine the magnitude of the response; the larger the
number of bound receptors, the more complex the response.

RECEPTORS

Every cell in the human body contains receptors, protein
molecules that act as docking stations for neuropeptides. Brain
cells, skin cells, eye cells—no matter what kind of cells we refer to—contain receptors, different types with differing functions, billions and billions of receptors.

Even with a limited set of receptors, quite complex behavior can be produced by the reception of a signal. . . . The cell could begin crawling, change direction, switch from a flat shape to a round ball, begin using up its resources, or begin creating new cell machinery. A cell possessing even a limited number of receptors is still simultaneously sensitive to all the signals that it is able to receive. Multiple signals may act together to produce responses that a single one would not generate.[6]

When you experience an emotional trigger, your brain produces neuropeptides. These peptides travel through your nervous and circulatory systems and attach to receptor sites on cells. If you were to experience anger or frustration, the peptides to help with the response for this situation is produced, travels, and locks on to the receptors of the cells involved. The cell then does what it needs to do to help with the situation, or to make the situation worse, depending on the type of peptide.

Situations demand responses. Neuropeptides can do nothing unless they seek-out and lock onto appropriate receptors. *Any old receptor will not do.* The receptor must have the same characteristics or *specificity* as does the neuropeptide that seeks to bind to it. A receptor with specificity A will not accept a neuropeptide with specificity B. It just will not happen. It is the same situation as when a guy goes courting a young lady; if she doesn't like what she sees, he doesn't have a chance! This is nothing but sexual attraction at the molecular level.

Receptors are embedded on the surface of dendrites across the gap in the synapse cleft. They are protein molecules embedded in the plasma membrane of a cell. In fact, receptors are located throughout the body, on the surface of every cell, on every organ, in every corner imaginable, and in any quantity uncountable.

Basically, receptors function as sensing molecules—scanners. Just as our eyes, ears, nose, tongue, fingers, and skin act as sense organs, so, too, do the receptors, only on a cellular level. They hover in the membranes of your cells, dancing and vibrating, waiting to pick up messages carried by other vibrating little creatures, also made out of amino acids, which come cruising along—through the fluids surrounding each cell. We like to describe these receptors as "keyholes," . . . constantly moving, dancing in a rhythmic, vibratory way.[7]

When a neuropeptide senses a receptor with the *same specificity*, it is like a marriage made in heaven. They fit together like a *lock and key*, having a natural affinity for each other. Once docked and fixed to the cell membrane, neuropeptides create a disturbance in the receptors. They vibrate; they wiggle; they dance; they change the shape of the cell until the information contained in the peptide enters the cell. A chain reaction of biochemical events then roar into action, producing a myriad of changes conducive to the *language* in the peptide, and, in the case of a new-born Christian—a holy matrimony!

IMMUNOPEPTIDES

Not only do neuropeptides occupy specific receptors capable of causing alterations in emotions and mood, but the nine immune components in the body, whose function is to attack invading infectious agents, pathogens, lock into the very same receptors! To add insult to injury (so to speak), so do viruses! Think of it—neuropeptides, immune cells, and viruses, all with the same specificity, roaming throughout the body, seeking whom they may attach to, to effect differing responses in the cell—neuropeptides to elicit euphoria, immunes to affect health, and viruses to make us feel bad. What a system!

All this sounds mind boggling—well, it is. There certainly exists a scenario that whatever is attracted to the receptor first, whatever is closest to, or in the immediately vicinity of, the receptor emitting a chemical scent of attraction, will win the race and lock on, preventing the others from gaining entrance to that particular receptor. From all this, it seems that an infinite number of possibilities exist within the cell where mixtures of euphoria, health, and disease may reside.

> The immune system was potentially capable of both sending information to the brain via immunopeptides and of receiving information from the brain via neuropeptides (which hooked up with the receptors on the immune cell surfaces).[8]

Neuropeptides and immunopeptides function in response to our thoughts, emotions, and attitudes in a system-wide feedback loop. They flow uninhibited. Neuropeptides produce emotions and emotions produce immunes.

> Neuropeptides of emotion carry information across the body-mind-spirit interface and back again. Thoughts of sadness, guilt, and despair tend to make us sick while thoughts of gladness, hope, and humor tend to make us healthy. The immune system is a mirror to life, responding to its happiness and anguish, its laughter and tears, its excitement and depression. Emotional experiences are inextricably connected to our state of health.[9]

CHAPTER THOUGHTS

The nervous system of the body; the brain and the spinal cord, the CNS and the PNS, constitute the main wiring diagrams of the body. Add to all this, those little critters traveling up and down the pathways; neuropeptides, immunes and viruses—and out comes a cellular communication system unparalleled in the history of humanity; dwarfing the creation of the stars and the planets, the Sun and the Moon, the mountains and the oceans, the seas and all that is in them.

Neurons and synapses, dendrites and axons, peptides and receptors, all come together to play in unison a musical symphony of life. Any wonder then that the human body can be looked upon as a massive information processing system, capable of producing euphoric pleasure or excruciating pain, abiding health or debilitating disease—either one having the propensity of rising above the others at the drop of a hat, producing varying degrees of the final result. Peer into, if you are able, the unimaginable tiny world of the atom and into the unfathomable complex world of the human genome.

TIDBITS OF PART II

- We are all—Carbon atoms, molecules, genes, and DNA:
  - o Like Amino acids A, T, C, and G.
  - o Like A bonds with T; C bonds with G—Always.
- DNA—Helix structure, right-hand twist:
  - o Like Carbon, Oxygen, Hydrogen;
    - ♦ Nitrogen and Phosphorus.
  - o Like transcription;
    - ♦ mRNA.
  - o Like translation;

- ♦ rRNA, tRNA, proteins.
  - o Like introns;
    - ♦ jDNA junk! Really?
- In the beginning God created Atom (I mean Adam!).
- Her name—Henrietta Lacks, died 1951, is alive today!
  - o Now wait: Is she alive or is she dead?
    - ♦ She's alive!
  - o Is she immortal?
    - ♦ Looks like it.
  - o Does she have a heart, liver, and kidneys?
    - ♦ No.
  - o How then is she alive?
    - ♦ She's alive at the cellular level.
    - ♦ Dead at the populational level.
- We are all:
  - o Cells– Cytoplasm, Nucleus, Nucleolus;
    - ♦ 100 trillion of them!
  - o Each with 35 thousand Genes!
    - ♦ Genome → book.
    - ♦ Genes → words.
    - ♦ Bases → letters.
- Nervous system:
  - o CNS, PNS, SNS, ANS, Sympathetic and Parasympathetic.
  - o Dendrites;
    - ♦ The receiver.
  - o Axons;
    - ♦ The transmitter.
  - o 100 billion neurons;
    - ♦ 100 billion synapses!
  - o Consisting of;
    - ♦ Neuropeptides.
    - ♦ Receptors.
    - ♦ Immunopeptides.
  - o They talk all day, all night;
    - ♦ Rain or shine.
- Information! Information! Information!
  - o Elicits emotions;
    - ♦ Feeling the information!
  - o A Symphony of Life!

PART III

Things We Believe

8

I WONDER:
DO YOU KNOW HIM?

"But whom say ye that I am?" (Mark 8:29)

Before we proceed with the topic before us in our discussion of the relationship between theology and biology that ultimately produces distinguishing *marks* of a Christian, we must firmly re-establish several monumental facts, assuredly cast in concrete, that bear upon our analysis, for if but one proves false, then all are futile.

Christians follow along behind the life as led by Christ. We claim to be the recipients of the benefits accomplished at the cross; of being born again by the power of the living God; to live a Spirit-filled life; and finally to be with Him and to enjoy Him forever. But there are four questions that must be firmly answered and believed in before we proceed, ere all is not well:

- Is Jesus God?
- Is Jesus human?
- Who is His father?
- Will He sit on the throne?

All four questions must be answered to the satisfaction of the Biblical record, for if Jesus is not God, then He is not qualified to die on the cross for my sins and yours; if He is not human, then He is a liar for claiming to be so, walking around ancient Israel as if He was, enticing everyone to look up to Him; and if He doesn't have a *human* father, well, then it really doesn't matter at all about where He is going to sit.

IS JESUS GOD?

THE TRINITY

It is firmly established in Scripture that Christianity stands or falls on the doctrine of the Trinity, the Triune Nature of God. We believe, and emphatically proclaim to all, that the God portrayed in the Old Testament is the very same God as portrayed in the New Testament, in relation to position, function, and as to number of Persons. The Old Testament God is a Trinity of Persons, as is the New Testament God the very same Trinity of Persons: God the Father, God the Son, and God the Holy Spirit.

The Trinity is that God is *One in essence* and *Three in Persons,* but that does not mean that the Trinity is composed of three different "gods," or as three different "parts" of one God. Nor should the term *Person* be understood to refer to a separate entity or being, for this would divide the divine essence. In complete terms, the Father is God, the Son is God, and the Holy Spirit is God; yet at the same time the Father is not the Son or the Spirit, the Son is not the Father or the Spirit, and the Holy Spirit is not the Father or the Son. Hard to understand? You bet it is. I still don't completely understand it all!

The knowledge of Jesus Christ as the Messiah comes with our understanding of the God of the Hebrew Scriptures. As we delve into the mysteries of His nature, and as the veil is lifted from our eyes in a progressive revelation, we come to an understanding of who He is as a God of personality and emotions. It is often difficult then to distinguish between the Father and the Son in the Old Testament as Christ was not yet incarnate as He is in the New. Many times when we read of God speaking in the O.T., it is Christ who is actually speaking:

> In the New Testament, Christ is clearly visible and distinct from the Father, while in the Old Testament, the Father and Son are often difficult to distinguish... In the New Testament, an *incarnate* Jesus Christ becomes visible to the world . . . while in the Old Testament, Christ also represents the Father – but not in *incarnate* form. However, in the Old Testament, when God appears, it is consistently the Son who appears, but as the representative of the Father, speaking the Father's words.[1]

THE SHEMA

In Judaism, the *Shema Yisrael* hangs above the very throne of God in majesty, in reverence, and in authority. Judaism rises or falls, depending upon every *jot and tittle* written in the Shema, to proclaim that the Hebrew God, the Tetragrammaton—YHWH of the Old Testament, is One in Nature, in Essence, and especially in Persons. No other doctrine in the entire Torah can hold a candle to the Shema, as it proclaims the very truth that God is One, and not three, nor is He somehow three gods in one, as Christians, they say, do believe.

The Shema is Judaism's declaration of faith, the "line drawn in the sand" concerning its view of monotheism—God is absolutely singular. Judaism affirms those religions that agree with the Shema, and vehemently rejects those that disagree with the Shema, especially Christianity with its Trinitarian doctrine of Three Persons in one Godhead. Christianity is the only religion on the face of the planet that affirms the doctrine of the Trinity as it relates to God. All other religions are either Unitarian as in Judaism and Islam; Polytheistic as in Hinduism and in many of the Asian religions; Pantheistic as in Buddhism; or Atheistic as in secular Humanism.

Christians believe in God as One, existing in Three Persons, a mystery no one can fully explain to the satisfaction of everyone else. However, the majority of evangelical Christians cannot easily defend the doctrine of the Trinity as contained in the Old Testament, against the onslaught of criticism mounted by orthodox Jewry and similar monotheistic religions. Therefore, for our discussion of the text, it would be advantageous to rewrite the Shema as it appears in the transliteration of Deuteronomy, and compare this with the English translation:

Shema, Yisrael: Yahweh Elohenu, Yahweh Ehad (Dt. 6:4).
Hear, O Israel: The LORD our God is one LORD.

Deuteronomy 6:4 is indelibly engraved upon the heart and mind of every devout Jew. They would, unhesitantly, bring to your attention clearly recognizable facts; that *the LORD our God* is contained in the Hebrew phrase—*Yahweh Elohenu,* and that—*is one LORD* is identical to *Yahweh Ehad,* as quoted above. Jewry would then have you to admit that Yahweh, the Hebrew name for God, is monotheistic in nature with no overtones of plurality, and would point to the phrase *"is one*

LORD" for undeniable evidence. But critical to the Christian is that the Hebrew God—*Yahweh* of the Old Testament, must be the very same God as the New Testament, Greek Trinitarian God—*Theos,* in order for O.T. Scriptures to flow seamlessly into N.T. Scriptures. Christians, who are not versed in Hebrew grammar, struggle for a satisfactory explanation as to why the Holy Spirit would include such a damaging verse in Scripture which seemingly destroys the Trinitarian doctrine.

It is recognized that the English word for God in the Genesis 1:1 account is actually the Hebrew *Elohim* and that Elohim is plural. Masculine plurality in Hebrew is identified by the *"im"* suffix. Therefore, *in the beginning God* [Elohim] *created the heavens and the earth.* In addition to this and upon closer inspection of the Shema, *Elohenu* is the possessive form of the name Elohim with the added suffix *"enu"* (our); Elohim is plural but Elohenu is singular! Are they both correct in identifying God as a singular *and* a plural deity?

Orthodox Jewry concedes that Elohim is plural, but they answer that the plurality does not identify God as a plurality of *persons,* but that it signifies that God exhibits a plurality of divine attributes in his many acts of creation (a sleight of man, not of Scripture). Added to this line of reasoning is their undeniable assertion in the Shema that Yahweh is *one* God, a singular pronoun that forever settles the question before them.

There must be something else, some convincing proof beyond all doubt that the Trinity is in both Elohim and Elohenu. No private interpretation of existing texts will qualify; no cleaver manipulation of pronouns or suffixes will suffice—something major, something perhaps in the Shema itself that will be an undeniable acceptance of Israel's true Savior, *Yeshua Hamashiach*—Jesus the Christ.

For the Christian, Jesus speaks definitively on the Trinity in Isaiah 48:16, *There am I: and now the Lord GOD, and his Spirit, hath sent me*—identifying three distinct persons—firstly, *I* and *me* (Jesus); secondly, *the Lord GOD* (the Father); and thirdly, *and his Spirit*—three Persons, not one!

EHAD VS. YAHID

As everyone knows who has tried, it is very difficult to try to proselytize a Jew into the Christian faith. The reverse is also true; not many Christians will renounce their faith to embrace the orthodox Jewish faith. Jews have their Unitarian God and Christians have their Trinitarian God, and never the twain shall meet.

But if there is any hope of swaying the Jew to consider their Messiah as the Second Person of the Trinity, if there is one way to break down *the middle wall of partition between us* (Ephesians 2:14), between Jew and Christian, it is to be found in the Shema itself. The Shema is the one textual passage that will convince the Jew of what the prophet Zechariah proclaimed centuries ago, that Israel had pierced to death their Messiah in unbelief, for Jesus Himself proclaimed: *And they shall look upon me whom they have pierced* (Zechariah 12:10). The apostle John, in the corollary to this in the Revelation, identifies Jesus as the Messiah using similar language:

> *Behold, he cometh with clouds; and every eye shall see him, and they also who pierced him* (Rev. 1:7).

There is another Hebrew word that is used over against the word *ehad*, as used in the Shema, for indicating the value or nuance of "one," that the Jew has never considered, and that has never entered into his mind. Convince the Jew of this and he will be forever indebted to you.

Consider Abraham when he was about to offer his son Isaac as a sacrifice to God on Mount Moriah:

> *And he said, Lay not thine hand upon the lad, neither do thou anything unto him; for now I know that thou fearest God, seeing thou hast not withheld thy son, thine only son from me* (Gen. 22:12).

Hidden within this verse is probably the word that has saved countless numbers of Jews, who have abandoned their monotheistic God and have embraced their Messiah in the Christian faith. The word is "only," Hebrew *yahid*, and means "one." This is Abraham's *only* son, his only *one* son, his only *absolute unity* son! There is no other son born to Abraham. Ishmael did not count as he was of Hagar and not of Sarah and, therefore, was not counted by God as a son.

Over against yahid, as used in the sacrifice of Isaac, stands *ehad* in the marriage ceremony:

Therefore shall a man leave his father and his mother, and shall cleave unto his wife: and they shall be one flesh (Gen. 2:24).

The first marriage between Adam and Eve in the Garden of Eden, two persons, male and female, came together in holy matrimony by God Himself. This union produced the first family consisting of man and woman, of husband and wife, both cleaving to each other and, thereby, becoming "one" in marriage, that is, by becoming "one" flesh. However, read carefully here; "one" here is *ehad*! It cannot be *yahid*! Yahid implies absolute unity, but there are *two* in this marriage. The author was obligated to use ehad, as it has the nuance of more than one, in this case two. There are two persons in this marriage and ehad correctly identifies that. Two persons in marriage are in the sight of God as "one"—*ehad, compound unity!*

The Jew is very confident in his daily recital of the Shema. He will defend it to the bitter end against anyone who would dare challenge the truthfulness of it. However, one obscure word, one word that has never entered into the mind of countless Jews, the word *ehad* could shatter his world forever:

Hear, O Israel: the LORD Our God is **ehad** LORD,

a *compound unity LORD*, not *yahid*, an absolute unity LORD! God does not exist as *a one, absolute unity deity!* God exists as a *one, compound unity deity!* The Jew is now in full retreat under the awesome weight and burden of his beloved Shema.

IS JESUS HUMAN?

Is Jesus God?—YES HE IS! But who is He? Is He human? Where does He come from, and of equal importance—who is His father?

We have demonstrated beyond doubt that He is God, but if He doesn't have a human father and mother as you and I do, then He can do nothing for you and me. Listen to me!—Jesus must be God *and* He must have human parents with the same type of genes that you and I have, ere we are all to be pitied!

Have you ever thought of the fallen angels that sinned along with Satan several millennia ago? These angels are now waiting until the end of this age to receive their just rewards:

And the angels which kept not their first estate, but left their own habitation, he hath reserved in everlasting chains under darkness unto the judgment of the great day (Jude 6).

These angels are seen as not belonging to the general classification of *elect angels* for Christ did not die for them after they sinned, as He died for us after we sinned. In order to apply the benefits of a salvation procured by Christ, He must be as *one of them!* There is no indication in all of recorded history that Christ was ever born as an angel; therefore, they are not offered salvation. To step into outer space—did Christ die for Androids? For Martians? If there are other creatures out there, how many times does He have to die? Once for each type?

On our planet, Christ must be human and human only! But the answer to this set of questions may startle and perplex some of you for years to come. Great minds throughout the annals of history have grappled with these questions. Untold scenarios have been advanced, some good, some very bad, but one thing is sure—we really don't know, with absolutely certainty, the answers to these perplexing questions.

It seems that every time we enter into a new discussion, particularly the one facing us now, we are led back to the early chapters in the Book of Genesis. I wonder why it is that way. Maybe it is because everything we have to say is rooted in those chapters. I believe that it is because of one man, the first man, Adam, and of whom he is the ancestor of, as all humanity *must originate in and from him* (Christ included).

Unless Adam was created first, *of the dust of the ground* (Genesis 2:7), and then Eve *taken from man* (v. 22), they being genetically the *same*, there could not be *one* Redeemer to stand in the place for *all* mankind. If Eve was created as Adam was created, *both* out of the dust of the ground, they being genetically *different*, there would then be three species of humanity. Adam would be of the first, Eve of the second, and first-born son, Cain, the progenitor of the third major species. Both Adam and Eve would have disobeyed God's commandment not to eat *"of the tree of the knowledge of good and evil"* (v. 17), *both* would have committed original sin; Cain and all his descendants would be living in sin as we do today. The only difference is that, in this scenario, *Christ could not possibly have been born sinless!* It is critical, then, that the male line of Adam be

the dominant factor in determining ancestral relationships as
we have today and not in a supposed line of Eve, she being genetically different: *For the man is not of the woman; but the
woman is of the man* (1 Corinthians 11:8).

LINEAGE, PEDIGREE, OR WHAT?

You and I both came from the loins of Adam. You and I,
and everyone else that will ever live, are in the loins of Adam.
We are all related to Adam. You and I are in his loins as pertaining to sin and death, in his natural seminal fluid. Jesus is
in his loins as pertaining to spirit and life, in his spiritual seminal fluid. Now some very thoughtful expositors of today would
say to you that the spiritual body of Jesus came directly from
God. This, however, is not correct. We cannot arbitrarily bypass Adam and then pray for a salvation! Jesus descended
from Adam as Luke proclaims in (3:23–38): *Jesus (v.
23)....which was the son of Adam, which was the son of God* (v.
38). It was Adam who came directly from God, not Jesus.

Have you ever thought of this—why is it that the Spirit
would take precious space to include several chapters on the
genealogy of Jesus, to list His father, His grandfather, His
great-grandfather, and on up or down the line. Was it because
He wanted to punish us, to make us meander endlessly
through a barrage of Biblical names that even their mothers
had difficulty in pronouncing, let alone spell correctly.

But Job (God bless Job!), gives us a glimpse of the problem
facing all humanity. We are all from Adam and therefore in
need of a Savior to come and redeem us from the penalty of sin;
a man who is as we are, and at the same time, also a God as
God is—essentially a God–man, a "daysman": *Neither is there
any daysman betwixt us, that might lay his hand upon us both*
(Job 9:33), a One that can touch us both. But where do we find
such a Person? Where should we start looking? I have never
seen such a One, never thought of such a One, and could never
recognize such a One. If there ever was such a One, it is of
immense proportions that we firmly establish who this One is;
where does He come from, and particularly, who is His father.

HIS GENEALOGY

To turn the pages of history ahead, the One that we are all
looking for is none other than Jesus Christ, the Son of God, the

Son of Man, the *daysman* that for centuries we have longed for. But does He satisfy the requirement of Job to be able, on the one hand, to understand the plight of man and his predicament, and on the other, to present Himself as a sacrifice to appease the wrath of God—to lay His hands upon us both.

A genealogy is a record of ancestors or descendants, depending on which way you look at it. The Bible provides two genealogies for Jesus, one in Matthew 1:1–17 and the other in Luke 3:23–38. Matthew's account moves forward in time, providing a list of descendants beginning from Abraham to David and finally to Jesus; whereas Luke's account moves backward in time, providing a list of ancestors starting at Jesus going back to David, then continues all the way back to Adam, and finally to God.

Matthew places emphasis on the royal descent of Jesus for he lists many of the Hebrew kings, especially David. One difference between the two genealogies is that Matthew mentions the line from Solomon to Jesus, whereas Luke mentions the line from Nathan to Jesus, another son of David but who was not promised the Davidic throne. Whatever these two genealogies of Jesus accomplish, whatever else may be said of the myriad of theological explanations put forth to account for apparent discrepancies between the two; one must indicate a link to David in order for Jesus to satisfy O.T. passages naming Him to be future King of Israel that sits on the Davidic throne:

> *Therefore being a prophet, and knowing that God had sworn with an oath to him* [David], *that of the fruit of his loins, according to the flesh, he would raise up Christ to sit on his throne* (Acts 2:30).

HER GENEOLOGY

And the other must establish the priestly line of Aaron. Assuming, rightfully, as many expositors do, that Luke is a listing of Mary's genealogy, she was then a descendant of King David through his son Nathan. Both Joseph and Mary then were descended from David. In Luke 1:5, Mary's record confirms her biological ancestry to David and then to Aaron, the High Priest of Israel. This establishes the perpetual throne, well into the Millennial Kingdom when Jesus comes again; to rule as King of kings and to reign as the Great High Priest in possession of an *endless life*:

*The LORD hath sworn, and will not repent, Thou art a
priest for ever after the order of Melchizedek* (Psa. 110:4).

But wait; hold-on now! God is seen here to inject a conundrum: What do you mean "after the order of Melchizedek"?
Who is this person Melchizedek? What does he have to do with
the priesthood of Jesus?

We all know that Jesus is not of the tribe of Levi, the
priestly line of which is Aaron, but from the tribe of Judah,
where no mention is made of priests coming from this line.
Now what? Is Jesus also prohibited from sitting on the throne
as well as not being able to officiate priestly functions?

By His sovereignty, God has circumvented the order of Aaron—an order that cannot continually perform the functions of
the priesthood; by installing the order of Melchizedek—one of
an endless life, one of whom it is said to live *forever*—without
descent, having neither beginning of days, nor end of life; as is
also Jesus; like unto the Son of God:

*Without father, without mother, without descent, hav-
ing neither beginning of days, nor end of life; but made
like unto the Son of God; abideth a priest continually*
(Heb 7:3).

Without elaborating in a myriad of details, Luke's genealogy firmly establishes Jesus' right to perform the functions of
the Great High Priest throughout eternity; whereas Matthew's
genealogy establishes Jesus' right to sit on the Davidic throne,
but not through his "father" Joseph. How then can Jesus establish his biological ancestry back to David?—no lineage back
to David, no right to sit and to rule on the throne. Jesus must
be of the human family, a *Homo sapiens*, as both you and I are,
with human father and with human mother.

BUT WHAT ABOUT CONIAH?

God has established the kingly line through Solomon, a
royal line that exists today that will resurface in the future
during the Millennial Kingdom:

*He hath chosen Solomon. . . . I will establish his king-
dom for ever, if he be constant to do my commandments
and my judgments* (1 Chr. 28:5, 7).

But this verse seems to be conditional—"*if he be constant to do my commandments and my judgments.*" But wouldn't you know it, there arose a descendant of Solomon named Jechonias: *And Josias begat Jechonias* (Matthew 1:11), commonly referred to as Jeconiah, alias Coniah. He was an evil king, so evil that God pronounced a curse on his descendants:

> *Thus saith the LORD, Write ye this man childless, a man that shall not prosper in his days: for no man of his seed shall prosper, sitting upon the throne of David, and ruling any more in Judah* (Jer. 22:30).

If Jesus is in the line of David through Solomon down through Coniah, is He now prohibited from sitting on the Davidic throne because of Coniah's sins? Are all the prophecies in the Bible stating that Jesus will sit on the throne of David shattered, for now no man in Coniah's line *will ever sit on the throne of David and rule.*

Not so fast! There is a little obscure note in Luke 3 that presents immense proportions. What don't you know about the little unnoticed phrase—"*as was supposed*"?

> *And Jesus himself began to be about thirty years of age, being (**as was supposed**) the son of Joseph* (v. 23).

"*As was supposed*" (or, everyone thought that) carries the connotation that Joseph was not the actual biological father of Jesus, for indeed he wasn't. So, again, what's going on here?

WAS HE ADOPTED?

If Joseph is not the biological father of Jesus, then is Jesus Joseph's stepson by adoption? Does this alleviate the obstacle in avoiding the curse of evil King Coniah when God promised that no descendant of Coniah would ever sit on the throne, Joseph also excluded, himself being a descendant of Coniah. However, there is no evidence in the N.T. that Joseph ever adopted Jesus as his son. Also, why should adoption be permissible since Joseph, being in the line of Coniah, is also prohibited from sitting? In addition, adoption rights to the throne did not exist in Israel. It is never recorded that an adopted son ever sat on the throne of David; therefore, He could not simply inherit the throne in this manner. Another requirement is that a future king must be a descendant of a *sitting* king and be appointed by that sitting king in order to be seated himself.

But the major obstacle against adoption is that David was promised a natural heir, a direct physical seed, and not an adopted son—an actual biological relative—a pure *son*.

WHO IS HIS FATHER?

Is Jesus God? YES! Is Jesus human? YES! This leads us directly to the "overshadowing" of Mary in His Incarnation, an act mysteriously performed by the Holy Spirit by method of *in vitro* insemination, the introduction of semen into the genital tract of Mary. From this point onward, the birth of Jesus was an absolute, normal birth.

> *The Holy Ghost shall come upon thee, and the power of the Highest shall overshadow thee* (Luke 1:35).

His was a birth of natural means by parents: Mary donating an X chromosome and the Holy Spirit donating a Y chromosome to produce a male offspring. This is the way it must be in order for Jesus to be identical to humanity. Any other method or procedure would declare Him to be non-human.

This conception is miraculous in the sense that the X chromosome donated by Mary consists solely of her *spiritual* ovum, and the Y chromosome deposited by the Holy Spirit contains *spiritual* spermatozoon obtained from His biological father. Our questioning now becomes more pointed: 1) from what human pool did the Holy Spirit obtain this infinitely valuable, biological, spiritual, Y chromosome, and 2) how come Mary's X chromosome is spiritual when that everyone born of man, that is, Adam, suffers from original sin in their genome?

WHEN GOD SPEAKS

If I could assume the position of the proverbial "fly on the wall," listening to the conversation between God and Satan; where Satan is revealing his desire to usurp total power from the Most High and thus to become God himself, assigning God to a lesser position, I would be astonished to hear God say:

> *For thou hast said in thine heart, I will ascend into heaven, I will exalt my throne above the stars of God: I*

will sit also upon the mount of the congregation, in the sides of the north: I will ascend above the heights of the clouds; I will be like the most High (Isa. 14:13–14).

At this point in history, Satan is confident of the fact that the promised birth of the Messiah can never become a reality, realizing that it is the Messiah that is said to be the One who will, at some future time, defeat him and cast him into hell. Satan here claims the victory; but God has other thoughts:

Yet thou shalt be brought down to hell, to the sides of the pit (v. 15).

"But how can this be," exclaims Satan, "This is impossible! The birth of your Messiah can never happen! How can a sinful spermatozoon come in contact with a sinful ovum in your chosen maiden, produce a sinless Messiah that will qualify Him to die for the sins of the world and send me to hell? Impossible!"

Or is it? God amuses and simply quotes another passage of Scripture to Satan that surely rocked his angelic foundations:

And I will put enmity between thee and the woman, and between thy seed and her seed (Gen. 3:15).

But how can this be? How can there be a conflict between thee (Satan) and the woman (Mary), and between thy seed (non–elect) and her (Mary's) seed—Christ: *And to thy seed, which is Christ* (Galatians 3:16). Satan is well aware that Eve was taken out of Adam. Her reproductive germ cells and organs were removed from Adam. He rightfully assumed that when Adam sinned, sin passed upon all men, including women. But he wrongfully assumed, to his ultimate demise, that Eve's germ cells were *also* tainted with sin. Or were they?

Even Bildad, a friend of Job in 25:4, is confused and asks the question: *How can he be clean that is born of a woman?* How can a woman come from a man as Eve came from Adam and not contain sin in her seminal fluid? It becomes apparent that God has designed both Adam and Eve to be able to circumvent this apparent contradiction. God answers by simply reaffirming that Christ will come from the "seed of the woman," where the seed, in His accounting, is sinless and immortal.

If Adam and Eve were blessed and bore a child *prior* to sin entering the world through the Fall, that child would be sinless and immortal. Therefore, it has been postulated that if a woman's egg, an ovum, containing sinless germ cells, when fertilized by a sinless sperm of a male, a spermatozoon, will produce a zygote with a sinless germ plasm—immortal.

But when Adam sinned, everything changed. So, how can Eve, sinful in her body, produce a sinless person as Jesus must be, to be able to place one hand on man, as the Son of Man, and the other hand on God, as the Son of God, to effect reconciliation betwixt God and man, as stated in Job 9:33?

THE GERM PLASM

It has been claimed by many of the ancient Jewish writers that Adam was created "with two faces," clearly indicating that Adam was originally bisexual with both male and female reproductive organs. When God created Eve, He did not use new "dust of the ground" as in Adam's creation. He took a "rib" from Adam and made Eve, meaning that He took the female entity out of Adam, leaving him with only male components.

The scientific theory of August Weismann[2], dubbed *"the Continuity of the Germ Plasm"* states that an individual germ plasm is derived *directly from the germ plasm of its preceding parents*. As figuratively depicted in Figure 8.1[3], Eve's germ plasm containing hereditary material always remains a germ plasm, never to originate out of the somatoplasm, i.e., the body cannot generate the seed but the seed generates itself *and* the body. Nothing that happens to the body's somatoplasm or body cells can influence the germ cells, i.e., *nothing in Eve's sinful somatoplasm body cells can transfer her sin nature to her sinless germ plasm cells* as received from Adam!

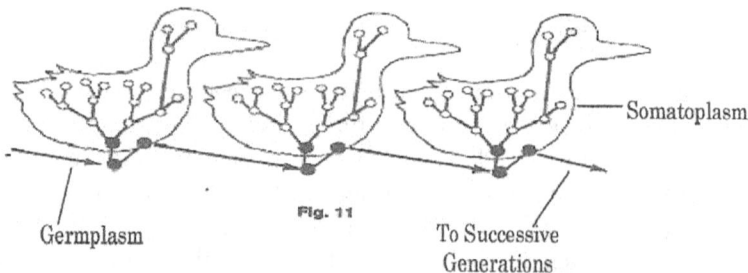

Figure 8.1: The Immortal Germ Plasm

Eve's body cells are disposable with its sin nature, but her female germ plasm is immortal with its sinless nature! Her body is subject to natural death, the germ cell to eternal life. Eve's germ plasm lives on, eventually releasing ova of itself into the gene pool that, when again fertilized by spermatozoa, will constitute a second generation of humans. This is the condition which Eve received from the hands of God in the Garden, sinless, immortal germ plasm cells—but not so with Adam. When Adam sinned his germ cells became corrupt by the sin mutagen (an agent that causes a change in genetic material), mortal and sinful, to be transmitted to all his offspring, as death came by man, not by woman. The only thing left standing, untainted by sin, was Eve's germ plasm!

Please allow me, your humble servant, at this juncture in our analysis of this difficult subject concerning our Lord, of the absolute necessity that He be human and possess a human father as you and I do; to elaborate and expand somewhat on the theory of Weismann, to record my personal understanding of the issues, and to offer some remarks:

• Adam transferred to Eve his *sinless,* female reproductive germ plasm *prior* to the Fall, allowing Eve, and ultimately all women, to remain *sinless* forever in her ova and to be the biological mother of our Lord.

• Adam passed on to all men his *sinful* seminal fluid, obtained *after* his Fall, from which all humanity came.

• Eve's germ plasm expressed high levels of the enzyme telomerase, enabling the cells to maintain the proper chromosomal length forever and therefore remain incorrupt, immortal.

• The cells of the somatoplasm in contrast, stopped producing telomerase, lost a portion of their chromosomes at each cell division and eventually died. They are corrupt, mortal.

• It now remains to be determined, from whom and by what procedure has the Spirit obtained this precious, sinless Y chromosome, seminal fluid at Christ's appointed time of birth.

Many have attempted to satisfy all that is involved in the "seed of the *woman*," in explaining many avenues of approach to bypass the curse of Coniah, some with apparent success, others with notable failures; but few have attempted to solve the insurmountable "overshadowing" of Mary by the Holy Spirit, by not addressing adequately "the seed of the *man*."

ENTER DAVID: STAGE RIGHT

We need now to find a *man* who is in possession of sinless spermatozoon to fertilize the sinless ovum of Mary, in order to foil the plans of Satan, and to enable humanity to become sons of God. A natural man cannot accomplish this as his seminal fluid is sinful, and would contaminate Mary's sinless ovum.

Now, what is meant when the Bible states that Jesus is *of the fruit of his* [David's] *loins, according to the flesh* (Acts 2:30), or that Jesus *was made of the seed of David according to the flesh* (Romans 1:3), or that Eve is said by Adam to be "bone *of my bones and flesh of my flesh*" when he gave up all vestiges of Eve in his own body? These terms are emphatic:

> *And Adam said, This is now bone of my bones, and flesh of my flesh: she shall be called Woman, because she was taken out of Man* (Gen. 2:23).

This is the only occurrence in the O.T. where it uses an additional prepositional phrase added to both direct objects—"bone" and "flesh", namely—"of my bones," and "of my flesh" as indicated in (v. 23). This surely indicates that Eve's bone and flesh were *directly* taken from Adam's bone and flesh, indicating a *direct descendant* of Adam; definitely not alluding to a descendant somewhere down the genealogical line of Adam.

> This phrase is a Hebrew mode of expression, denoting emphasis and certainty. Such reduplication is a vehement affirmation, partaking of the nature of an oath: and where such is used, it was that men might know God is in earnest in that which He expressed. It also respects and extends the thing promised or threatened: I will do without fail, without measure, and eternally without end.[4]

Out of all humanity in recorded history, in order for Christ to be human, the deposited sperm (hold-on now!) must come from *David*. This condition can only be satisfied if David is deceased and in heaven in possession of his spiritual body (in whole or in part); thus *bypassing Adam's sinful sperm*. David's spiritual body is still human, albeit refined, but is now sinless, immortal, and not subject to death. Now that is some spermatozoon!

Also, what does it mean that God took a "rib" (reproductive organs) from Adam to create Eve? If He created Eve in that manner, why can't He take a "rib" (seminal fluid) from David to create Jesus in the same manner? What, in all our sacred Biblical theology, has been compromised?—all that there is in the "overshadowing" is still there, nothing being omitted.

We have, therefore, revised the genealogy chart of Jesus, Figure 8.2, to account for the foregoing discussions; placing Him directly under His biological father David, a *direct descendant*; removing Him from Joseph's line; of whom it is certain that he is not the biological father of Jesus (*as was supposed*). God used 23 chromosomes from mother Mary and 23 from father David to produce a Homo sapiens—JESUS!

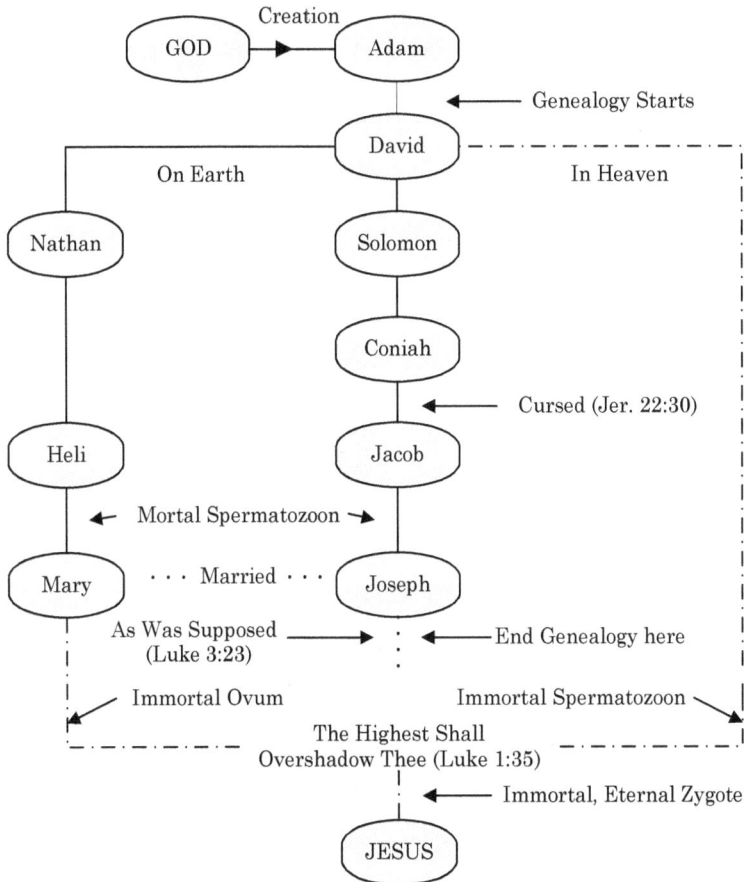

Figure 8.2: The Revised Genealogy

Do you really think that there are no *active* saints in heaven today; David in particular? Really? Well, think on this:

> *But ye are come unto mount Sion, and unto the city of the living God, the heavenly Jerusalem, and to an innumerable company of angels, To the general assembly and church of the firstborn, which are written in heaven, and to God the Judge of all, and to the spirits of just men made perfect, And to Jesus the mediator of the new covenant* (Heb.12:22–23).

This passage ranks among the most difficult to interpret in all of Scripture. Without certainty and in good faith, expositors have applied this verse to whatever they desire it to mean. But, taking it at face value, it speaks directly to our statement that David is in heaven today complete in his spiritual body. Notice the following that are listed as *now living* in heaven, called Mount Sion (Zion), the heavenly Jerusalem:

1. God lives there.
2. Angels live there.
3. The church of the firstborn lives there.
4. Jesus lives there.
5. Just men made perfect live there.

Just men made perfect, including O. T. saints! *David!* They are alive, just, active, and in the presence of the Lord Jesus.

This, in essence, is the miracle of the overshadowing!

WILL HE SIT ON THE THRONE?

Is Jesus now eligible to assume the position as King of Kings and Great High Priest in the Millennial Kingdom, to sit on the throne of David as promised in the O.T.? Is He in the *"bloodline"* of evil king Coniah, thus exempt from the throne?

> *And when thy [David's] days be fulfilled, and thou shalt sleep with thy fathers, I will set up thy seed [Christ] after thee, which shall proceed out of thy bowels, and I will establish his kingdom. He [Solomon] shall build an house for my name, and I will stablish the throne of his [Christ's] kingdom for ever* (2 Sam. 7:12–13).

We answer with an emphatic *NO*, for His bloodline *ends at David*. His titular or royal line continues through Solomon and is firmly established in *bypassing* the curse of Coniah by method of "Levirate Marriage," a type of marriage in which the brother of a deceased man is obligated to marry his brother's widow. Briefly, both Matthew and Luke list the name of Salathiel (alias Shealtiel) and Zorobabel (alias Zerubbabel), where *Jechonias begat Salathiel, and Salathiel begat Zorobabel in* (Matthew 1:12); and in Luke 3:27, *Zorobabel, which was the son of Salathiel, which was the son of Neri*.

Now, while in captivity in Babylon, Coniah (in the royal line of Solomon) married a daughter of Neri (in the priestly line of Nathan). His daughter was widowed but she had a son named Pedaiah. By levirate law, Coniah now has the privilege of claiming Pedaiah as his son. Later in life, Coniah fathered a son through Neri's daughter and named him Shealtiel. This action makes Pedaiah and Shealtiel brothers.

To make matters worse(?), Shealtiel, married by this time (don't know her name), died childless, which says that incumbent upon Pedaiah that he now marry the wife of his brother Shealtiel, and later produced a child named Zerubbabel who is the last king listed in the genealogies. But Zerubbabel (actually Shealtiel's *step-son* in our economy) is in the bloodline of Nathan, not Solomon, whereas Shealthiel was the last in the bloodline of Solomon. The curse on Coniah runs true—no sitting king in Solomon's line after Coniah.

CHAPTER THOUGHTS

The Jew can now view his Hebrew Scriptures as both authoritative and reliable as it is written; that in Genesis 1:1, *Elohim created the heavens and the earth;* that in 1:2, mention of the *Spirit of God* is made in the Hebrew word *ruach*, wind; that in 1:3, mention is made in the first day of *Let there be light,* when there was no sun created until day four, alluding to Jesus as *I am light of the world* (John 8:12); and in Genesis 1:26 in using the plural pronouns *us* and *our: And God* [Elohim] *said, Let us make man in our image, after our likeness*. Yahweh is One and Elohim is Three; both are true!

The identity of the male sperm in the conception of Jesus will remain a topic of discussion forever, as well it should. This should inspire awe and gratitude in our hearts for who He is and of what He has accomplished.

Among all humanity, Jesus is the only One who satisfies Job's request for a *daysman* who can faithfully reconcile sinful man to a righteous and holy God.

The Holy Spirit *overshadowed* Mary in the conception with spiritual spermatozoon. We have introduced the concept that it was the seed of David, King of Israel, for it is stated that Jesus is the Son of David, *of the fruit of his loins, according to the flesh* (Acts 2.30). He is the biological, bloodline son of David. His genealogy *ends there*, thus is not involved in the curse of Coniah; qualified; claiming the right to sit on the throne of David in the Millennial Kingdom—King of Kings.

The womb of Mary is not simply an incubator with no active participation in the formation process of Jesus. Though her blood was tainted with sin and could not cross into the placenta to pollute her fetus, nevertheless her blood flow was the medium by which filtered nutrients entered the embryo of Jesus and the cause of His continued growth. He is truly both of "the seed of the woman", and "the seed of the man (David)."

So also the Holy Spirit "overshadows" our nucleotide structures to transform the natural to the spiritual that eventually results in the birth of an elect Christian. There can be no biological difference between His human nature and body, and our human nature and body. The difference lies in the fact that our new spiritual nature derives solely from His spiritual nature, as His *incorruptible seed* is in us (1 Peter 1:23).

Matthew 1:1 is of paramount importance, for it is emphatically stated in no uncertain terms that Jesus is the son of David. He is never referred to as the son of Solomon or of anyone else. In the future millennium kingdom when Jesus sits on His throne, will He not testify that: *I, Jesus....testify unto you these things in the churches. I am the root and the offspring of David.* (Revelation 22:16)? The bottom line is this: God "overshadowed" Mary with David's spiritual spermatozoon! The miracle is in His "overshadowing." His birth was normal.

What a blessing! Now we may be confident that our sins are forgiven; that we are sons of God; that we are the temple of the Holy Spirit; that we cannot sin in our spiritual nature; that we will be resurrected at the rapture of the Church; that we are the Bride of Christ; that we will come with Him to rule and to reign in the Millennial Kingdom; and that we will judge both the wicked and the dead at the end of the age.

Hallelujah, what a Savior!

9

MURDERER! WHO? *ME?*

"Thou shalt do no murder" (Matthew 19:18)

G od has accomplished great things in predestination—in the creation of the universe, of who Christ is, of His divine and spiritual natures; of His birth into the human family and of His genealogy. His absolute credentials are essential in order to sacrifice and take away sins, and hence of His death and resurrection. This leads us to affirm that Jesus is the Christ, the Son of God, the Son of Man, who will one day sit on the throne of King David in the millennial kingdom to judge all the families of the earth.

He created all of us in a manner that we call "biological." There is no biology in the stars, in the galaxies, or even in the moon. Why would anyone ever want to go there, let alone live there? The only place in the universe where we could survive biologically is on *terra firma.* He has done this for us and in the meantime, we have done absolutely nothing for Him!

But we may object to this and relate to everyone all the many things we did to put us in good standing with the Lord; of how we prayed, went to church, sang in the choir, and lived the good life that enabled Him to choose us and thus to be called Christian. These are all noble gestures but, sad to say, these are not the prerequisites for being called Christian. It behooves us then to consider what *truly marks us out as Christian,* for God is not yet through performing great things.

What was it that qualified us for such a coveted position? When was it that we claimed this honor; for the first observable mark did not occur during our regeneration in transforming from a nonbeliever to believer; nor was it at a time earlier in life at our natural birth when we knew very little of Christ;

but it actually occurred *centuries before our time,* during a momentous occasion in an unbelievable scenario.

Believe it or not, it all has to do with *murder!* It has to do with the relationship that Abraham had with his great-grandson Levi (son of Jacob and Leah), but then resurfaces centuries later at the time of the death of Christ. Listen closely to what the Scriptures say concerning this drama, a blessing bestowed on all those who, from that time forward, will eventually believe in the Savior, confirming the Sovereignty of God.

THE CAST

Jesus Christ, Lord and Savior to all who believe, came to this earth to accomplish one specific purpose in life; to suffer on a Roman gibbet and die for sinners. He left His home in heaven and emptied Himself of His Glory at birth; born of a Virgin. We beheld Him as a child, teaching truth and righteousness, instructing the rabbis in Old Testament prophesies that spoke of Him. He was despised and rejected of men, a man of sorrows and acquainted with grief. He suffered *as it were great drops of blood falling down to the ground* (Luke 22:44) in the Garden of Gethsemane, for what was in the Cup for Him to bear. He gave up His spirit and arose victoriously; seated at His right hand—our King and Great High Priest, forever.

Volumes of literature that fill countless libraries around the world have expounded on the life, death, burial, and resurrection of Jesus Christ. But there is one chapter that has never been written, one chapter that has never entered into the hearts and minds of men. The question now raised before us, as pertaining to his death and which has never been seriously addressed, nor adequately answered, is—*Who killed Jesus?*

Was it the Roman legions that nailed Him to the cross under the direction of Pontius Pilate, the Governor of Judea, responsible for the death of Christ? Were all the Jews who yelled: *Crucify him, crucify him* (Luke 23:21), fostering centuries of anti-Semitism, responsible? Or, is it all humanity who will ever live the culprits who killed Him, each and every one of us, Christian and non-Christian alike?

Many theories abound as to who is, or who are, the villains who perpetrated this evil deed: to slaughter the Divine Son of God; and by so doing, erroneously thought to nip Christianity in the bud; to prevent countless millions of people from enjoying an eternal life with Him. Each scenario is played-out again and again on the stage of human history.

TWO MEN OF OLD

Although there is a great deal of truth in what expositors have written of the facts in the death of Jesus, none have advanced beyond what the Bible summarily seems to be saying concerning the events leading up to His death. To answer the question adequately, references must be made to the theology of the Old Testament, and there is none better, nor more qualified, than the apostle Paul, to draw out and systematize the passages that bear on this most important subject, paramount for our very salvation.

The Apostle develops the doctrine that there are *two representative men* in history, Adam and Christ. Adam is the *first man* and Christ is the *last Adam* (1 Corinthians 15:45). Adam represents all humanity because of original sin, and Christ represents all the elect because of His atoning work on the cross. All humanity is in Adam because of imputed sin and transmitted sin nature; all the elect are in Christ because of His substitutionary death. Imputed sin comes to us directly from Adam; the transmitted sin nature comes to us directly from our natural father (the "T" in TULIPS, Total Depravity).

In Genesis, the prohibition to not eat *of the tree of the knowledge of good and evil* was given to Adam (Genesis 2:17), not to Eve. Though Eve ate of the fruit first, she did not sin first, for she was not prohibited from eating of it. Adam then ate of the fruit and he sinned, for he was directly prohibited to eat of it. All mankind sinned immediately after Adam sinned, including Eve:

> *Wherefore, as by one man sin entered into the world, and death by sin; and so death passed upon all men, for that all have sinned* (Rom. 5:12).

Adam was the first person to sin and death was his to bear. But because of Adam's sin and the pronouncement of the judgment of death upon him, both sin and death passed on to all men. But, you may object to this, in that, you may say, you were not even there: How could I have sinned if I wasn't there?

The Apostle is aware of your objections, and writes a few lines of explanation that establishes the doctrine forever.

In effect, Paul says that the first prohibition by God was given to Adam; the next prohibition was given in the law at Mt. Sinai through Moses to a rebellious people. There was *no law* or prohibition by God against sinning during the period stretching *from* Adam *to* Moses. Sin was still in the world, but sin was *not imputed* to man; it was not charged to his account:

> *For until the law sin was in the world: but sin is not imputed when there is no law. Nevertheless death reigned from Adam to Moses, even over them that had not sinned* (Rom. 5:13–14).

Nevertheless, men died during the period from Adam to Moses. The Apostle asks the question: "Why did men die when there was no sin imputed to them?" They did not sin as Adam sinned by disobeying the direct prohibition by God, therefore, why did they die? If they sinned, and they did sin, but were under no such law not to sin, then, why did they die?

A modern example of the doctrine of original sin could be explained by the birth of a new born baby. Unfortunately, the baby is born, takes one breath, and then dies. Surely the baby did not have a chance to sin, did not know how to sin, and could not even think of what sin was. Yet, the baby died. Why? The answer rings loud and clear; because Adam is our *representative man*, and we all sinned *in him* when he sinned because we are all in the *loins* of Adam.

Now wait, you may ask, what does all this mean? The writer to the Hebrews takes up the charge, speaking about Levi paying tithes to Melchizedek, King of Jerusalem, while *yet unborn* but in his father Abraham's *loins*, writes:

> *And as I may so say, Levi also, who receiveth tithes, payed tithes in Abraham. For he was yet in the loins of his father, when Melchizedek met him* (Heb. 7:9–10).

God has determined, in the genome makeup of man, that it is possible for someone *yet unborn*, to be present in generations *prior* to his physical birth, to actually take responsibility for some current events. Levi, yet unborn, paid tithes to Melchizedek, through the loins of Abraham generations before his own birth. Abraham here is a sort of representative man for Levi. On a larger scale, Adam is *the* representative man for all mankind, and we sinned in him. Therefore, all men living from Adam to Moses died because of their imputed sin nature, and not because of any personal sins that they may have committed during this period.

But we have not been left helpless: Christ also is a representative man because He *died once*, as Adam *sinned once*. If Adam did not sin once for all humanity, then Christ could not die once for all humanity. If this were the case, it would have been necessary for Him to die over and over again, once for every individual person, to affect his or her personal salvation.

The doctrine of original sin, therefore, is both necessary and sufficient: necessary because without it there is no real offer of salvation; sufficient because it satisfies the requirements of God that He die once for the sins of men. All humanity without exception *was in the Garden*, in the loins of Adam, when he sinned, and all the elect, without exception, *were at the cross at Calvary*, in the loins of Christ, when He died! There is no other way.

TWO NEW MEN

It was necessary to labor somewhat in order to establish the irrefutable fact that there are two representative men in history, Adam and Christ. In order to explain adequately the events leading up to the death of Christ, we now propose to introduce two *new* representative men, and attempt to demonstrate and justify their existence. They are classified as quasi-representative in the sense that they stand for us at a certain *time and place* in history, but not as Adam and Christ are representative with men in their loins for all time.

The first new representative man is placed under Adam, and he represents those who are *passively responsible* for the death of Christ, i.e., all those who did *not physically* kill Christ.

The second new representative man is placed under Christ, and he represents all those who are *actually responsible* for the death of Christ, i.e., all those who *physically* killed Christ.

Those who have been convicted of being passive in His death can never be accused of being actual, for they are innocent of the fatal blow that took His life. Both representative men were present at the cross at the time of Christ's death. All humanity must therefore be catalogued as either passively responsible for His death or actually responsible. It is here acknowledged that the Bible indicts all humanity without exception for the death of Christ, from Pilate to the Jews, and finally to all Gentiles. Nevertheless, we maintain that there are some who were passive and some who were actual.

THE PLOT

THE DIVINE SIDE
Every theological question and answer pair must consider two sides to be effective and complete; the divine side and the human side. On the divine side, the Old Testament sacrificial system was a *type* of the New Testament sacrificial system, the *antitype*. The Old Testament sacrificial lamb was a type of the final New Testament sacrificial lamb, Christ, the antitype.

The High Priest officiated and continually offered blood sacrifices to God for the sins of the people; whereas the Great High Priest, Christ, offered Himself to God as a blood sacrifice once for all for the sins of His people. Therefore, the High Priest is a type for God; the Great High Priest is God the antitype. It is well documented Biblically that God planned every detail in the death of Jesus:

> *Surely he hath borne our griefs, and carried our sorrows: yet we did esteem him stricken, smitten of God, and afflicted* (Isa. 53:4).

Strong[1] defines *smitten* (Heb, *nakah*) as "to kill, slaughter, or to slay," and of *stricken* (*nega'*) as "to strike violently, to destroy." The Trinity, *including* the Son of God, participated equally in the death of the humanity of Jesus. As on the Day of Atonement, the High Priest killed the sacrifice for the sins of the people, so also in the anti-type, God killed Christ at the Cross for the sins of His people. God is then passive and actual in the death of Christ; passive because it was in eternity past that the Trinity formulated the plan wherein Christ should die; actual because God *struck and smote him* in history.

THE HUMAN SIDE
In Christian theology, every believer is a priest. Adam was a priest; he offered sacrifices to God, Abel also after him. Eve-

ry member of a family is a family priest and is encouraged to sacrifice. Non-believers are not priests; they do not offer blood sacrifices. They brought fruit from the ground that is not acceptable to God, e.g., Cain. When a family priest offered a sacrifice in the Old Testament, there was a prescribed method of doing so, as typified in the sacrifice of the burnt offering:

> *If any man of you bring an offering unto the LORD. . . .*
> *let him offer a male without blemish: . . . And he shall*
> *put his hand upon the head of the burnt offering; and it*
> *shall be accepted for him to make atonement for him.*
> *And* he *shall kill the bullock before the LORD* (Lev.
> 1:2–5).

Leviticus is devoted to the worship of the redeemed people of God, and it is the sin offering, performed by Aaron the High Priest in Lev. 16, that secures salvation through the atoning work of Jesus Christ. But in the burnt offering sited above, it is the *offerer* himself who kills the sacrifice, not the High Priest. The action of the believer, in the offering of the lamb as a blood sacrifice, is seen as a vicarious and atoning sacrifice for the sins of the offerer. The laying on of the hands signifies acceptance and identification of himself with his offering; and in figure of the anti-type, it answers to the Christian's faith in identifying himself with Christ. The point to notice is that *only believers* in the O.T. economy sacrificed a burnt offering!

Non-believers are nowhere in sight with respect to the prescribed O.T. offerings. These two groups, the believers and the non-believers, the offerers and the non-offerers, are types that are found in New Testament anti-types. Both groups are found intermingling around the tabernacle in the O.T. sacrificial system, and both their anti-types are to be found intermingling around the sacrificial cross at Calvary in the N.T. system.

CRUCIFY HIM!

SCENE ONE: AT THE CROSS

Every dispensation in history has its altar. The Old Testament had several altars spanning thousands of years of history. The tribulation period will have an altar and the 1,000 years millennial reign of Christ will also have its altar. During the Church age, Christians also have an altar, the Cross; for it is said that Jesus willingly went to the altar of sacrifice. Now the events recorded in the crucifixion of Jesus are paramount to the understanding of who killed Jesus.

We start with the Gospels of John and Mark. Although the Bible is true and accurate in all areas of concern, it may not be totally chronological in order. Therefore, we have listed the verses of interest to us below and have indicated at what verses we rearranged to present a more accurate sequence of events, while maintaining strict theological accuracy.

The scene is at the Cross where Christ and the two male-factors are being crucified:

- **John 19**
  v. 31: *The Jews . . . besought Pilate that their legs might be broken, and that they might be taken away.*
  v. 34: *But one of the soldiers with a spear pierced his side, and forthwith came there out blood and water.*
  v. 30b: *It is finished: and he bowed his head, and gave up the ghost.*
  v. 33b: [They] *saw that he was dead already, they brake not his legs.*
- **Mark 15**
  v. 39b: *He* [the centurion] *said, Truly this man was the Son of God.*
- **John 19**
  v. 37b: *They shall look on him whom they pierced.*

Jesus was hanging on the altar and it was getting very late in the afternoon. The Sabbath was almost upon them and Jesus was not yet dead, and everyone wanted to go home to prepare for the Sabbath meal. We may summarize this scenario as follows:

- The Jews went to Pilate to ask permission to break his legs to hasten his death. The Roman crucifixion was a death of suffocation. By breaking his legs, Jesus would not be able to push down against his nailed feet, thereby lifting himself up to relieve his chest muscles in order to continue breathing; death would be imminent.
- But *one* of the soldiers (i.e. a *certain* soldier; a centurion) pierced his side with a spear and immediately blood and water

came out. Jesus was still alive at this point in time, His heart still beating in order for blood and water to immediately gush out. Most probably the legionnaire drove his lance between the ribs, upward through the pericardium and either pierced his heart or a main artery that could account for the details. The act of piercing His side with a spear was the instrument used in his physical death:

• After the piercing, it was all over; *the blood was shed!* He bowed his head and gave up His Spirit, and died.
• They saw that He was now dead; therefore, they had no reason to break his legs to hasten His death.
• The Centurion, watching all that was going on, admitted that this indeed was the Son of God.
• A predictive prophecy: *They shall look on him whom they pierced.*

SCENE TWO: HE'S COMING AGAIN

In the Revelation, when Jesus comes in the second part of his Second Coming (the first part was His coming in the Rapture preceding this event), the apostle John writes:

Behold, he cometh with clouds; and every eye shall see him, and they also which pierced him: and all kindreds of the earth shall wail because of him (Rev. 1:7).

In this scene, Christ is seen as coming to the earth to end this evil world system, and is going to divide the elect from the non-elect before the Millennium era begins. As He is seen descending from the clouds to earth, *every eye shall see him*, i.e., every eye without exception, to include both the elect and the non-elect then living on the earth. The writer addresses first the elect: *And every eye shall see him, and they also which pierced him!* Of all the statements that could ever be made at such a stupendous event, statements expressing His love and mercy, of His forgiveness and eternal life, of heavenly bliss and rewards in glory; of all these, He intentionally mentions the *piercing of His side!* And He mentions the piercing of His side in the plural, *they*, whereas the actual piercing done at the cross was done in the *singular*, by *one* Centurion! If a singular Centurion pierced to death Jesus, how then can the apostle John speak of *they* pierced to death Jesus?

Only one explanation is possible and true to Scripture: John 19:34 addresses the *one* Centurion who actually killed Jesus by piercing His side which establishes the fact that he is the *Representative Man.* Revelation 1:7 addresses the elect who also actually killed Jesus by piercing his side, and are identified with that sacrifice through the slaughter performed by the Centurion. The conclusions drawn from this are plain:

- The Centurion is the new representative man placed under Christ.
- The Centurion killed Jesus, and is *actually* responsible for his death, and he represents all the elect.
- All the elect *actually* killed Jesus.

The following is the refrain to that grand old song that has thrilled the hearts of every Christian throughout the entire world—words describing the crucifixion events at the cross, the memorable words spoken by Pontius Pilate, the governor of the Roman province of Judea—*I Find No Fault In Him* (John 19:4, 6):

> *Then they pierced Him so deep in His side,*
> *until the blood came streaming down.*
> *And that's how Jesus purchased my salvation!*
> *And I find no fault . . .*
> *I find no fault . . .*
> *I find no fault . . . in Him!*

SCENE THREE: THE OTHER GROUP
Unlike the elect who pierced the side of Christ to death as a blood sacrifice, the non-elect also pierced Christ, but this piercing did not result in a blood sacrifice. Jesus could die only for those who shed his blood for God demands a blood sacrifice.

The apostle John then goes on to address the non-elect in Revelation 1:7: *And every eye shall see him. . . . and all the kindreds of the earth shall wail because of him.* They are here seen wailing, or beating their breasts, for they now realize that they have rejected their Messiah and now face an impending doom. Zechariah adds more detail to the two groups addressed in the Revelation passage, of those who pierced and of those who mourn or wail:

- **Of those who pierced**

And I will pour upon the house of David . . . the spirit of grace and of supplications: and they shall look upon me whom they have pierced.

- **And of those who mourn**

And I will pour upon . . . the inhabitants of Jerusalem, . . . and they shall mourn for him, as one mourneth for his only son, and shall be in bitterness for him, as one that is in bitterness for his firstborn (Zech. 12:10).

Grace for those who pierce, and for those who mourn, bitterness. They have not exercised faith in Christ, but sight. King David speaks to the bitterness of the mourners and their participation at the cross as roaring lions and bulls:

Many bulls have compassed me: strong bulls of Bashan have beset me round. They gaped upon me with their mouths, as a ravening and a roaring lion (Psa. 22:12–13).

The priests, elders, scribes, Pharisees, rulers, and captains bellowed round the cross like wild cattle, fed in the fat and solitary pastures of Bashan, full of strength and fury; they stamped and foamed around the innocent One, and longed to gore him to death with their cruelties. Conceive of the Lord Jesus as a helpless, unarmed, naked man, cast into the midst of a herd of infuriated wild bulls. They were brutal as bulls, many, and strong, and the Rejected One was all alone, and bound naked to the tree.... Like roaring lions they howled out their fury, and longed to tear the Saviour in pieces, as wild beasts raven over their prey.[2]

The Psalmist goes on to identify those around the cross as *dogs*, the wicked; they who pierced the Lord's hands and feet:

For dogs have compassed me: the assembly of the wicked have enclosed me: they pierced my hands and my feet (Psa. 22:16).

Here he marks the more ignoble crowd, who, while less strong than their brutal leaders, were not less ferocious, for there they were howling and barking like unclean and

hungry dogs....Such a picture is before us. In the centre
stands, not a panting stag, but a bleeding, fainting man,
and around him are the enraged and unpitying wretches
who have hounded him to his doom.[3]

Sad as this commentary is of the natural man, his participation at the cross did not qualify as a blood sacrifice. Piercing
of the hands and feet do not have the connotation of killing as
in the piercing of his side, but only in the sense of wounding.
Simply hanging on a cross and expiring after many hours of
suffering have elapsed, because of nails protruding from his
hands and feet, does not qualify as a blood sacrifice.

The *dogs* (plural), alluded to are all the non-elect, and it is
said that they pierced, not to his death, but his hands and feet
to the cross. In doing so, they did not offer a blood sacrifice;
therefore they did not kill Christ. They are *passively* responsible for his death, not *actually* responsible. The passage then
goes on to identify the representative of the dogs:

Deliver my soul from the sword; my darling from the
power of the dog (Psa. 22:20).

We now know who the second representative man is; he is
the *dog* (singular)! The centurion pierced His side, but the dog
wielded the hammer and nails to pierce the hands and feet of
Jesus. Whoever he is, he represents the non-elect; the dogs,
who are passive in the death of Christ. We may now claim that:

• The *dog* is the new representative man placed under
Adam.
• The *dog* did not kill Jesus, but is passively responsi-
ble for his death, and he represents all the non-elect.
• All the non-elect, the *dogs*, passively killed Jesus.

It is not probable that the identification of the *dog* will ever be determined from the pages of Scripture, nor is it necessary that we know of his identity (though I am persuaded that
the dog is a soldier indwelt by Satan). What is necessary to
know is what he has done and whom he represents.

We may, however, draw some conclusions from Scripture
that would seem to point to a certain type of individual; a type
of individual that no other person can qualify; a person unique

in the entire world at that moment in history, which event could never again be repeated. In the Hebrew text, the word *power* is used 120 times, eleven of which uses the word *yad* which has the nuance of *absolute* power. The remaining 109 renderings use other Hebrew words for the same English word and signify nuances other than absolute power.

Nowhere in Scripture does it ever intimate that Jesus Christ should ever fear the power of any created *human being!* Then why stress power? What type of person could ever yield a degree of absolute power over Christ in his humanity? Satan?

ONE FINAL ATTEMPT

It is sometimes observed that Satan uses the non-elect to accomplish his primary desire to annihilate the Jewish people, to thwart the eternal purposes of God in salvation: *When the sons of God came in unto the daughters of men, and they bare children to them* in Genesis 6:4, was an attempt to pollute the Jewish race with intermarriages; of angels inhabiting human men, and then marrying human women. The offspring of the union, the *Nephilim*, the Giants, increased throughout the land. It was an attempt to prohibit the coming of Christ.

The result of this was a global cataclysmic flood that annihilated the human race, save Noah and his family to continue the Messianic line. The power behind the Nephilim to corrupt the race, as great as it was, could not withstand the onslaught of the power behind the flood, for God *cast them down to hell, and delivered them into chains of darkness* (2 Peter 2:4).

Then, there was Haman in the Book of Esther (3:6ff), who tried to annihilate the Jews by some diabolical plot to hang on the gallows all the Jews in the Babylonian exile, with the same expected result of interrupting the Messianic line. But God intervened by hanging Haman instead of the Jews.

Satan uses both men and animals to indwell in order to accomplish his intentions: the serpent in the Garden of Eden; demon possession of a man in the synagogue; Legion in the untamed man; the Beast in the Revelation; and the most striking of all, the temptation of Christ by Satan himself; in order to win over Christ by offering to Him all that there is in the world, thereby avoiding the cross. But this last futile attempt by Satan, prior to the cross, convinces him that his worldly

rule is about to be terminated, for he knows full well that at
the cross God said:

And I will put enmity between thee and the woman, and
between thy seed and her seed; it shall bruise thy head,
and thou shalt bruise his heel (Gen. 3:15).

Because of Satan's rebellion and his intrusion into the affairs of men, God has passed judgment on him—Christ shall
bruise his head; and here, 4,000 year later at the cross, the
prophecy was being fulfilled. Christ delivered a fatal blow to
Satan; spiritually, he was stripped of his power to rule the
world, and physically, in the near future, he will be cast into
the Lake of Fire never again to deceive the nations.

Satan *shalt bruise his heel,* by piercing his hands and feet,
before Christ bruises the head of Satan, for it is at His death
that the bruising of the head is prophetically fulfilled. But can
it be? Can it be that Satan, not willing to accept defeat by the
Son of God, fully realizing that this may be his last opportunity to destroy the Messianic promises made down through the
annals of history, knowing that his future is determined for
destruction, indwelled and energized the *dog* to perform his
appointed duty? Can it be that: *and thou shalt bruise his heel*
and, *they pierced my hands and feet,* are both prophesies of the
same kind and order; the first being a predictive prophesy and
the latter being the fulfillment of the *same* prophesy; the
bruising of the heel of Christ in Genesis 3:15 with the piercing
of His feet in Psalms 22:16! Has Genesis 3:15 been fulfilled to
the very letter, acted out in vivid detail over 4,000 years later
at the cross at Calvary?

The scenario is sure. Satan's attempt to interrupt the Messianic fulfillment resulted in complete failure; for not being
ordained to offer a blood sacrifice for his own sins, he offered a
non-bloody sacrifice of bruising and piercing His hands and
feet, in hopes that the soldiers would immediately break His
legs; and hence He would expire quickly as a non-bloody sacrifice. But God, who demands a blood sacrifice by his people, in
His infinite wisdom and power, raised up a certain Centurion,
causing him to pierce His side before He could expire from the
wounds of a non-bloody sacrifice. Believers are thus blessed
beyond measure: We have been given the honor and the privilege of sacrificing, yea, of killing the precious Lamb of God!

THE JUDGE AND THE JURY

Down through the annals of history, the Roman crucifixion was to cause extreme suffering upon the one hanging on the gibbet, to suffer there as long as possible. There is no indication recorded that the Romans exercised compassion upon the accused; no record of ever ending the life artificially by casting a spear into the side of the victim; and that under penalty of death. But God had other plans. He overruled the Romans! He wounded Him for our transgressions; bruised Him for our iniquities, *and with His stripes we are healed* (Isaiah 53:5).

How close all humanity has come to being eternally separated from the life of God, one seemingly obscure act of nailing, an act performed at every Roman crucifixion, but holding eternal consequences for humanity with respect to the Son of God. If Christ be not raised as a blood sacrifice, there is no possibility of Christianity and we all remain in our sins!

But we, who pierced His side to the death, sacrificed the Lord Jesus as a blood sacrifice acceptable to God. Those who pierced His hands and feet will never live again to have an opportunity to right the wrong. In jurisprudence, persons that are accused in a crime of killing are convicted of first-degree murder, while their accomplices in the same crime are convicted of second-degree murder. In the eyes of the law, all of the accused are guilty of the killing, but each has been judged differently for their level of participation:

• The non-elect have been charged, found guilty, and have been convicted of second-degree murder in the death of Jesus Christ, by association, and are indeed passively responsible for the crime.
• The elect have been charged, found guilty, and have been convicted of first-degree murder in His death, by premeditation before the foundations of the world, and are actually responsible for the crime.

Divine justice has been meted-out upon the human family:

• The non-elect, those not convicted of the actual killing, but only of association with the crime, were sentenced to—*Eternal Death!*

• The first time in history of jurisprudence that the elect, those having been convicted of the actual killing in the crime, were sentenced to—*Eternal Life!*

Do you recall that grand old song, the African-American Spiritual—*Were You There?*

> *Were you there when they crucified my Lord?*
> *Were you there when they crucified my Lord?*
> *Oh . . . .*
> *Sometimes it causes me to tremble,*
> *Tremble,*
> *Tremble,*
> *Were you there when they crucified my Lord?*

Yes, We Were There! Emphatically!

CHAPTER THOUGHTS

God is Sovereign in the universe! He will do whatsoever He will do. At one point in history, He wrote: *Thou shalt do no murder* (Matthew 19:18), but in eternity past He ordained and constrained you and me to slay the Lamb of God. I admit that I cannot fathom the depths of such a divine mystery. But one thing I know, and am absolutely certain of—if Christ be not sacrificed for the sins of His people, by His people, then heaven is a myth, hell is real, and the human race is forever lost.

Everyone that will ever live and breathe was in the Garden of Eden, present in the loins of Adam.

Everyone that will ever live and breathe was present in the Garden at Calvary—either represented by the *dogs* and present in the loins of Satan, the *dog*—or represented by the Centurion and present in the loins of Christ, our Savior.

Every elect Christian that will ever live and breathe is a *murderer* from the beginning, *effectually called* to be one, destined to eternally sing praises unto the slain Lamb of God!

Every unbeliever, every unprofitable servant, will have his eternal existence in the outer darkness: *And cast ye the unprofitable servant into outer darkness: there shall be weeping and gnashing of teeth* (Matthew 25:30).

10

BIOLOGICAL REGENERATION

"According to his mercy he saved us, by the washing of regeneration, and renewing of the Holy Ghost" (Titus 3:5)

L et's now leave behind those principles we've been discussing and forge ahead to the subject at hand. As it was in the original creation when God first created man, so it is now that God the Father, the First Person of the Trinity, now determined to be the *External Source of Empowering Energy*; further identified as the *Energizer*; the Holy Mutagen; the critical source of energy determined by the first requirement of the Second Law of Thermodynamics for suspension and reversal; invades, restructures, and activates the nucleotide sequences of the proto-predestination genes; impressing changes upon existing DNA structures to cause a change or mutation to a higher level of order; from a structure of death to a structure of life; mutations that work together in concert to bring about a final beneficial result. The Second Law of Thermodynamics, the inviolable law of sin and decay, is *suspended!—reversed!*—due to His Holy intrusion. *Hallelujah!*

THE BUTTERFLY—AN ANALOGY!

Everyone loves the Butterfly, but not everyone is aware of the regenerative miracles that take place every day in its life cycle. When the caterpillar has eaten enough in its larval stage (it is a voracious eater), it turns into a pupa and initiates the pupa stage of life. To accomplish this, all that is needed for him to do is to stop eating and find somewhere to safely hide for a while. The first thing that happens in this stage is that most, *not all*, of the caterpillar's old body dies, being attacked by the very same fluids that the caterpillar used when normally digesting its food. However, *not all* the tissue is destroyed; some of the insect's *old tissue passes on to become part of its new self!* These special tissue cells (does this sound to you like predestination cells?) have lain dormant in the caterpillar and have played no active role during the larval life; but has now become active to supervise the building of a new body out of the soup that the digestive fluids have made. In this process, old DNA genes have been "turned off" and new DNA genes, which have never been used before, have been "turned on." A new body has been brought into existence: the Butterfly!

We now define the butterfly to consist of two kinds of flesh—one kind that existed in the old larval body and can be assigned to natural caterpillar flesh, cDNA (c for caterpillar); and a new kind of flesh, bDNA (b for butterfly), existing now as a new entity—a butterfly (*not* an extension of caterpillar).

To be perfectly clear, the butterfly consists scientifically of two kinds of flesh, one flesh is of the remnant of the original caterpillar flesh and the other flesh is of the new metamorphosed butterfly flesh. How much cDNA flesh and how much bDNA flesh constitutes a butterfly, I do not know (you will have to ask a Lepidopterist!).

ONE BODY, TWO NATURES

The body of Adam with its original, natural DNA flesh, emanating from the dust of the ground, is clearly seen as the initial step that God used in the creation of man in the Garden of Eden account; breathed into his nostrils the breath of life (Heb. *hayyah*); and man became a living soul. The plural for hayyah, *life*, is hayyim, *lives*, and is the *actual* grammar used in the Genesis account. There was *more than one* type of life created in Adam; a natural life of the dust from which he was created, and a spiritual life endowed with the image of God. With the first two steps in the creation process completed, instantly, man became a living soul. But the *fall* of Adam into sin changed all that. Adam lost his spiritual life; the *image of God* departed; the Holy Spirit no longer resided in his body.

Now fast forward to the present time. Gleaning from the discussion thus far, I have come to the following understanding of both theology and biology; each of us, each human being, is born with a *natural* life with *nDNA flesh* (n for natural), and a *propensity* to obtain *spiritual* life with *sDNA flesh* (s for spiritual). This can only come into existence by a direct, divine intervention in the originally created natural DNA flesh mutated by the action of the deadly carcinogenic agent of sin.

METAMORPHOSIS
And be not conformed to this world: but be ye transformed by the renewing of your mind (Rom. 12:2).

The Greek, *metamorphoo*, is translated in our Romans passage as a *transformation*, from which we get our English word *metamorphosis*. Webster defines the term as:

A transformation, as by supernatural means; to change from one form to another . . . a transformation of one kind of tissue into another.[1]

Metamorphosis—*a transformation of one kind of tissue into another*. WOW! The Christian body is changed from the inside out by the action of metamorphosis, similar in many ways to the metamorphosis of the caterpillar; the total remaking of a new species never before seen. This is nothing less than the performance of a divine miracle! This should be a constant reminder to all of us, that when speaking of the miracle of metamorphosis of the human body, from the natural to the spiritual, we should remain aware that we are treading on holy ground, and possess no language to adequately explain the process. But one thing is certain: God's intention is to change that which is natural to that which is spiritual; from *one kind of flesh into another kind of flesh!*

The transformation of a butterfly is a beautiful picture of the actions taken in the metamorphosis of an elect Christian. The Holy Mutagen performed an act of supernatural transformation in the nature of human flesh, pouring untold amounts of energy into an otherwise dead system to affect an intended response. A new creature has been brought into existence, not unlike the miracle in the creation event of Adam; or in the making of water into new wine at the feast in Cana of Galilee; of Him walking on water; or giving sight to the blind beggar. These occurrences are divine miracles that, of necessity, suspend the actions of the Second Law.

ONE BODY

Variations in the Genesis created kinds are *horizontal*, not vertical. Horizontal in that God performed no new creation in going from, say, Collie to Beagle. If variations in this "dog kind" were vertical, then we would have a new kind or species from that of Beagle, no longer a Beagle of the dog kind.

There is currently no new creation-kind going on in the world today. The Second Law prohibits vertical variation within the Genesis kind, as vertical variations to occur would have to have an influx of new information to construct the new kind. This is against the Second Law, which is universal in scope that tends to go downward to a lower complexity, to a loss of information; unless constrained otherwise by some predetermined, external programs and mechanisms.

The effects of election constitute a new creation, resulting in a new kind! God introduced new information into existing molecular and cellular structures consisting of nDNA to cause a vertical variation in the original created "man" kind. This intrusion resulted in the transformation of nDNA into sDNA, from the natural to the spiritual; from the natural man to the spiritual man; an upward mobility from that which was in existence before. Humanity now has two flavors: one—the non-elect with nDNA flesh, *mortal;* and the other—the elect with sDNA flesh, *immortal;* a higher, more complex form of human nature—a Christian (*no longer* an unbeliever).

An elect Christian, by virtue of being impregnated with the spiritual sDNA of God, is not a mere variation within a kind but a new form of human being with the highest level of applied information within any of God's creations. God has performed a new miracle; a new kind. No scientific process can account for this new kind, no microscopic differentials can observe changes in its structures, nor be observed in its patterns.

Let's face it—we've been changed! *Hallelujah!*

TWO NATURES

The glory of the old image, belonging to the *old man,* is now broken down, dissolved, and done away with. It is then created into a new form of glory, that of the glory of the *new man.* We are constantly being changed, from one fading glory into another magnificent glory, from corruption unto incorruption, from mortal unto immortal. What God is after is simply a new man, a *new creature,* where all elements at the molecular level be changed from one form of flesh into another

The unique relationship of two natures in one body produces a soul not seen before in the annals of humankind. A singular nDNA body and soul expresses its characteristics in a solitary fashion; one attribute of personality, one of emotion, and one of will. But the body and soul with two independent, autonomous natures, one of nNature and one of sNature, as in that of a Christian, expresses its distinctive characteristics in a *dual,* but independent fashion; as two attributes of personality, two of emotion, and as two of will. Both natures act independently of each other but act as one solitary person with one consciousness; producing a plethora of seemingly contradictory activities, and at times, seemingly complimentary.

Now we know assuredly that:

Cancer Genes of sin are in our old nature. Predestination Genes of life are in our new nature. Both natures play-out in the human body! Both are there! In the biological realm, Cancer and Death, and Election and Life, are played-out in, and determined by, the human genome!

The appearance of sDNA in the body is the start of the spiritual man, a nature that has been created *for* heaven that is to reside *in* heaven in the future life. It is to be continually molded by the action of metamorphosis into a sanctified body during the lifetime of the Christian. What spiritual values that reside in the sDNA Christian are there, containing all the attributes of the mature spiritual man in heaven, but on an incomplete level; still in the making, yearning to be complete. How much nDNA flesh and how much sDNA flesh constitute a Christian, I do not know (you will have to ask the Lord!).

THE BIOLOGY OF GOD

Now before you all fall off the edge of your chair, I do not mean that God is to be known and studied by what we humans know and have studied about human biology, and then extend that to deity. What I do mean is that God interacts with the human body via the biology of the human body! If He doesn't act upon my biology, I will never know that He is there. I will never hear Him, understand Him, nor ever pray to Him. If His thoughts for me, or His presence within me, do not seek out and lock onto receptors within my body that possess the same *specificity* as His thoughts and His presence to cause its desired reactions, then I am theologically blind, biologically inactive, and on my way to a Christ-*less* eternity.

After all, the Greek *bio* means "life," and with *logia,* "the study of," tacked onto the end of it, means the study of life. Surely, God is life—why not then attempt to study that life and attempt to make a few preliminary opinions. After all, if Jesus walks the way I do, if He talks the way I do, and if He eats the way I do, then He is the way I am!

All three members of the Trinity are thought-*full* Persons. When each thinks, and those thoughts interact with the human body, it must, of necessity, react with the thought processes and mechanisms of its biological counterpart in the human body. Divine thoughts must seek out the specific biological receptors that will be conducive to react to those thoughts. The thoughts will then dock onto those receptors; the receptors will wiggle and dance to transfer the meaning of the divine message into the cytoplasm of the cell. From there it may go to—who knows where!—but one thing is sure, it will enter the nucleus of the cell to perform the function in the DNA structure that it was intended and designed to perform.

We will extend this analogy to the divine thought of predestination that finds its biological corollary in human regeneration.

SDNA OF JESUS

The spiritual body of a Christian is paralleled by, but not identical to, the human body of Christ, wherein resides two natures, not one of the natural and spiritual as we, but one of deity and one of spiritual. Only the spiritual form of the body of Christ would be comparable as He is the pattern out of which the *new man* is created. We may gain valuable insights into the workings and nature of the sDNA body when we look at the life, ministry, and resurrection of Jesus, both during the time of His ministry on earth and also after his resurrection.

I propose to you that those who looked upon Jesus during His lifetime, regardless of their moral classification, saw a man with a totally spiritual body in every respect, without blemish and without sin. Therefore, a totally spiritual sDNA body is not a gas or some type of ether, but can be looked upon and viewed even with nDNA eyes. The question then remains: Is an nDNA body that we may see any different in appearance from that of an sDNA body? The answer to that question is a resounding no, but inwardly there is a difference as wide and as deep as the gulf that exists between the East and the West.

The natural body is made manifest by its actions in the world, but the spiritual portion of the body of a Christian, as indeed present within the natural body, cannot, as yet, be discerned. We cannot differentiate between our natural and spiritual bodies with our physical eyes. We cannot divide the essence of our being into its dual parts. No one knows what our sDNA body looks like when we are looked upon, but it must be equal in composition to Christ's spiritual body.

This analysis may be extended to His earthly body as there is no Biblical indication that the appearance or composition of His earthly body changed after His resurrection. The apostle Thomas confirmed in the upper room that what he was looking at was indeed Jesus resurrected from the dead, in possession of the same recognizable, spiritual body of sDNA flesh as before His death. The earthly body of Jesus was the very same body that He was seen with as He rose up from the earth and then into heaven itself, to forever exist as the God–Man:

> *Ye men of Galilee, why stand ye gazing up into heaven? this same Jesus, which is taken up from you into heaven, shall so come in like manner as ye have seen him go into heaven* (Acts 1:11).

Jesus possessed two natures, one nature of deity, emanating from His essence as God, and one spiritual nature emanating from His sDNA humanity. His spiritual nature is no different from your or my spiritual nature. His divine nature in no way interfered with His human nature, both acting independently of, and in harmony with, each other; but at the same time supporting one another; acting as one individual person with one consciousness. This independency of natures allowed His spiritual body to be totally human and equal in every way to the spiritual body of every Christian. When Jesus said: *"I am the way, the truth, and the life: no man cometh unto the father, but by me"* (John 14:6), His divine nature was in play, not his spiritual nature. His sNature was inactive in this context and could not perform this function, as only deity is able to accomplish this feat. Against this, Jesus said: *"I thirst"* (John 19:28). Here Jesus is thirsting in his sNature, as His divine nature is not capable of thirsting. Both natures are seen to act independently of each other, as He has no conscious thought of which nature is acting at any one time. He acts as normally as you and I would act, after all, He's human also.

The spiritual body of Jesus was entirely sanctified whereas in our body only sDNA is totally sanctified; the nDNA body is partially sanctified, "set apart," still in the making, waiting for the time of its sure regeneration. Our two natures do not, in any sense of the word, live in harmony with each other, as do the two natures of Jesus. Our natural flesh acts according to our natural nature and our spiritual flesh acts according to our spiritual nature.

Our sDNA body can do everything that the body of Jesus could do, even walk on water if it ever was to be loosed from its moorings to our nDNA body. We would be able to go through doors and to appear at will as He did. To be spiritual, then, is to assert that the Holy Spirit abides in the spiritual man, in the molecular structure itself, in our sDNA flesh.

THE VIRUS

To obtain an *earthly* picture in order to see and to understand the intricate spiritual workings of each of the members of the Trinity, we need only to look at the goings-about of the common influenza virus; and that by using a similar analogy of the male sperm *relentlessly* seeking to penetrate the female oocyte to produce a new life.

A beautiful example of what *may be happening* within a Christian body during the process of regeneration; of how the Father, as the External Source of Empowering Energy, the critical source determined by the first requirement of the Second Law of Thermodynamics for suspension and reversal, orders the metamorphosis of natural cells infected with sin, into spiritual cells infected with the image of God.

As we start, have you ever considered that you are never alone! You and the world that surrounds you are covered with viruses and bacteria. They are the ultimate survivors. Some are good; some are bad. Viruses belong to the smallest of things, ranging from barely 10 nanometers and up to 300 nm in size. They come in different sizes and shapes. They are classified by scientists as both living and non-living (possibly somewhere between the two), depending on who you talk to. They may seem and act like living organisms because of their unending ability to reproduce, but viruses are not living organisms in the strictest sense of the word.

A virus is not a bacterium, a single-celled organism. In fact, they have no cellular structure as a bacterium has. They have no hearts, no livers, and no brains. They do, however, have nucleotide bases in either DNA or RNA molecular configurations, either double or single stranded, that allow them to make copies of themselves. They are, however, very intelligent. Viruses have only a few genes; for example, the influenza virus with only eight, the rotavirus with eleven, whereas in humans there are upwards of 30,000.

A virus is not an independently living organism. It must seek-out a living cell in order to exist and have any meaning in its life cycle. In a viral attack, the virus must first enter the host cell, quickly becoming an uninvited guest, and once there, to set up house and present its own viral DNA (vDNA or vRNA, v for viral) to the host nDNA cell. The virus does not have enough information in its meager nucleotide sequence to allow it to perform its intended function, which is to generate viral parts to later assemble them into new, mature, infectious viruses. They then exit the host cell to infect other host cells, to repeat the process over and over again.

Viral populations do not grow through cell division because they are a-cellular. They cannot synthesize proteins, because they lack ribosome and must use the ribosome of their host cells to translate viral vRNA into viral proteins. Instead, they exploit the host's organelles and commandeer the machinery of a host cell to produce multiple copies of themselves. Viruses cannot generate or store energy but have to derive their energy from the host cell. They are truly intercellular parasites.

ORIGINAL SIN

The virus may also give us an indication of the method Satan used when he infused sin into the genome of Adam, allowing for the normal biological functioning of the human body.

God created Adam without sin but he was commanded not to eat of "*the tree of the knowledge of good and evil.*" Sin was not in the commandment but was lurking within the "fruit" of the tree and when Adam *ate* of it, he sinned: *"for in the day that thou eatest thereof thou shalt surely die"* (Genesis 2:17).

What was it that caused Adam to sin and eventually die? Since it wasn't the commandment, then it must be the fruit or something in the fruit by taking a bite out of it and swallowing it. Physical death now began to reign in the human race— Adam and all of us now began to die!

After all this happened, God expelled both Adam and Eve and then locked-down the Garden so that Adam, and all humanity after him, could not reenter and then eat *"of the tree of life"* (Genesis 3:24) and live forever. This constitutes the doctrine of "original sin," as no one but Adam can hold this distinction. We sin *in* Adam, but not originally *as* Adam sinned.

Something was in the "good and evil fruit" that gave *knowledge* of good and evil to Adam. Knowledge is information. The question now remains: How was this information of evil transferred to Adam's biology, more specifically, to his DNA. Now I will be the first to admit that we will never know the answer to this question, but my vote is for an original *antigen/pathogen complex of sin* located in the *evil part* of the good and evil fruit! The original pathogen contained all the sin imaginable and is not recoverable today due to being locked-out. To be more specific than that, I cannot, but a few helpful definitions may be in order:

• **Antigen:** An antigen is not an organism and therefore not specifically biological, but is a protein molecule that works to our benefit at times by triggering a reaction from the immune system of the body, thereby eliciting an abundance of antibodies to fight against intruders. Antigens are often found on the surfaces of pathogens and are unique to that specific pathogen (like a pathogen's license plate).

• **Pathogen:** Microscopic organisms of bacteria, viruses, toxins, or ***other foreign substances*** (infectious agents) that cause sickness and disease by releasing deadly toxins into the infected host cell.

• **Antibody:** An antibody, also known as Immunoglobulin, is a protein molecule used by the immune system to identify and neutralize foreign objects like bacteria and viruses, i.e., antigen/pathogen complexes, by locking-on to the antigen and presenting a piece of it to other immune system components for further evaluation.

When the *original* antigen/pathogen complex of sin entered Adam through one of the normal pathways of his body, the mouth, the first thing that it met was B cells spewing-out all differing kinds of antibodies, billion of them. Normally, when encountering a pathogen for the *first time* in its life-cycle, B cells, with no previous encounters to store in its memory, did not have the proper antibodies to identify and to lock-on to the invading sin antigen in order to begin the process of destroying the pathogen complex. Having no mechanism to defend the genome, the original sin complex attached itself onto the membrane of a body cell, as in Figure 10.1², transferred it's lethal dose of toxic sin personified into the cytoplasm, and the rest is history. As copies of the original pathogen "budded-out" from the invaded cell into the bloodstream, Adam's entire body became infected with original sin, including his spermatozoa.

Adam passed sin on to all his succeeding generations, spreading throughout the genome of man as would a biological plague, but it was not the same as the original sin complex. The original copies in Adam's sperm, containing all manner of sin located in one individual pathogen, seem to have mutated into a host of specific pathogens with but one specific sin identity and an antigen specific to that sin. All sin is there, but not in one pathogen as in the original but in many individual sin pathogens. In this manner, when we commit a sin, it may be due to a specific sin pathogen with an antigen specific to that particular sin (e.g., the sin of hatred, or the sin of envy).

ETERNAL LIFE REGAINED

In an attempt to explain the initial act of regeneration, in transforming a sin-laden nDNA cell into a sinless sDNA cell, the Father implements the act of sovereign election by targeting predestination genes that lay dormant in an nDNA cell of an elect host—solely an act of Sovereign Grace for there is no "free will" at the molecular level.

Using similar terminology as in the metaphor of the influenza virus, *instead* of the virus, we *substitute* in its place the Father's divine Thoughts, dThoughts (d for Divine), penetrating the human body through the frontal cortex of the brain, producing divine neuropeptides, dNeuropeptides, with the very *same specificity* as His dThoughts. These dNeuropeptides roam throughout the body in search of cells that have been marked-out for salvation; and when located, dock-on to cell surface receptors, dReceptors, of identical specificity (can you now identify several biological systems in the figure below that we have been talking about all along?).

The original Message of God, now residing within the dNeuropeptide, a message of salvation to all who believe, then transfers its life-transforming instructions through the plasma membrane into the cytoplasm of the cell as shown (step 1); penetrates the nuclear envelope to "turn on" or "splice into" proto-predestination genes at specific sites on the host nDNA (step 2); impressing upon the host genome the image of God, thereby removing the sin pathogen out of the nucleotide sequences, effecting the *initial steps* in regeneration.

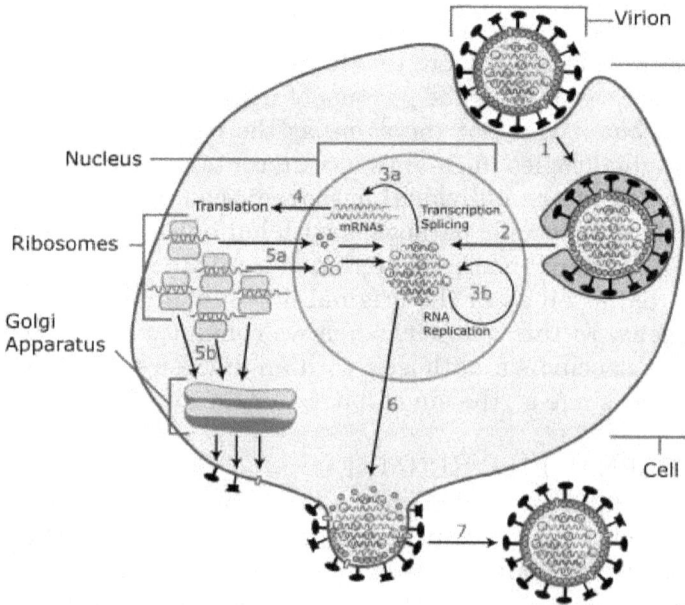

Figure 10.1: Virus Attack

Jesus Christ, the Second Person of the Trinity, the *External Source of Conceptual Information*; further identified as the *Informer;* the critical source of information determined by the second requirement of the Second Law of Thermodynamics for suspension and reversal; restores the necessary generic codes that God originally intended man to possess into the revised nucleotide sequences; enters the nucleolus of the cell and takes over the cell's replicatory functions; then cloning of the revised sDNA sequence into spiritual mRNA commences by transcription (step 3a) and replication (step 3b); where smRNA exits the nucleus and migrates into the cytoplasm (step 4).

The Holy Spirit, the Third Person of the Trinity, the *External Source of Conversion*; further identified as the *Regenerator*; the critical source of metamorphosis determined by the third requirement of the Second Law; *translates* smRNA into spiritual amino acids and proteins, the building blocks of the new spiritual man, by means of the host ribosome—to be transported back into the nucleus (step 5a) to undergo the assembly of all the spiritual parts, and through the golgi apparatus—to the plasma membrane (step 5b), where assembly of receptors occurs to allow a means of peptide release from the cell. With both initial acts completed, the natural nDNA cell is regenerated into a spiritual sDNA cell containing the complete image of God, nothing being omitted.

Images of the original dNeuropeptide containing the message of God are then budded out of the nucleus into the cytoplasm (step 6), and out through the plasma membrane, secreting copies of itself into the existing world of nDNA cells (step 7), free to roam throughout the body; to await the next signal to invade nCells earmarked for election. Host antibodies have been made ineffective, being duped to recognize the invading *new man* as a self-antigen that presents no threat to the host, causing no alarm to be given to the host nImmune system to launch an all-out attack upon the spiritual invaders.

Sanctification commences as original dNeuropeptides, budding out from sCells, begin to seek-out additional nCells that the Holy Spirit may direct for regeneration. The spiritual rule may be this: First the initial act of regeneration of the first nCells to sCells by the Holy Spirit, establishing a level of *positional* sanctification in the remainder of nCells; then regeneration of additional nCells to sCells by the continual budding-out operations of newly formed dNeuropeptides; bringing more and more cells to Christ; raising the level to *progressive* sanctification in the diminished remainder of nCells; and on and on it goes until death or the resurrection ends the process and ushers in *perfect* sanctification in all sCells.

Have you noticed that when the Father entered the cell to attack and rearrange the nucleotide sequence and to abolish the embedded code, that the sin-laden cell died! In the act of regeneration, cells must first die, as: *It is appointed unto men once to die* (Hebrews 9:27). But all is not lost. After the Son embedded new information into the DNA sequence of the cell for it to function in a spiritual atmosphere, the Holy Spirit performed the conversion process, and now the same cell that died has come alive again and will now live eternally.

CHAPTER THOUGHTS

God desires that all men be saved. He constantly issues forth His Thoughts, the call to confess our sins; to accept Jesus as our Savior. His Thoughts go out to all men without exception; enters into the body, producing dNeuropeptides containing the same message as in His dThought; scans up and down the spiritual pathways of the body, to seek-out receptors of the same specificity; finding an abundance within the elect, but, sad to say, finding none within the non-elect.

With this inflammatory mixture of sDNA and nDNA cells in the one body of a Christian, there exists a *war* between good and evil; of incorruption against corruption; of righteousness against unrighteousness; between the two rivals within the human genome; the two natures vying for *supremacy* of the one body. Ultimately, it is a battle between belief and unbelief, between the belief of the elect in the God of heaven and the belief of the non-elect in the god of evolution. There is no other choice, thus insuring that the battle will never end until death overtakes us. The non-elect with only nDNA flesh can never obtain an sDNA body; their predestination genes will never be turned on and will strive against all adversity to maintain their coveted position of unbelief.

Over against this is the body of an elect Christian. The sDNA body can never sin, is never in need of forgiveness, and can never die. The nDNA body always sins. Both bodies are at war with each other and the spiritual will always attempt to evangelize the natural, still a part of the elect body, but as yet has not been called-out nor transformed by the action of metamorphosis. This occurs when normal nCells are regenerated to spiritual sCells. We will not be able to entirely transform all cells in this life but God has the power to complete the process at the resurrection of the Christian body. All the while, sanctification of the natural continues. Confusing? You bet it is! But praise the Lord for His process of metamorphosis!

	Present Time		←|→ Eternity
ity ←|→			Future
	Phase 1 | **Phase 2** | **Phase 3** |	**Phase 4** | **Phase 5**	
	Babes | Children | Young Men |	Fathers | Sons	
tion	Zygote | Blastocyst | Embryo |	Fetus | New Birth	
	Molecular| Cellular | Organismal |	Populational| Adoption	
all			
egeneration	Justification		Glorification
	Reconciliation	Sanctification	

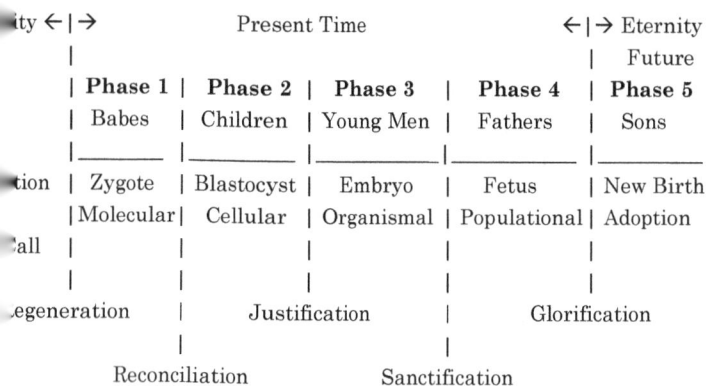

Figure 11.1: Spiritual Childbirth

SPIRITUAL CHILDBIRTH: PHASE 1

logians universally confirm that regeneration is the
the new birth zygote phase, and that the beginning of
tion ends the last phase of fetus. It is not to be implied
conciliation, justification, and sanctification need to
specific lengths of time until the believer progresses
several phases of growth before they become effective.
e understood that these three doctrinal classifications
ur simultaneously upon the act of regeneration and
re glorification. It is intended to associate a specific
to a specific phase of growth for clarity.

ENERATION

re there ever was a flower, a rain drop, or a snowflake
om the sky; before there ever was an earth, a moon, or
the sky; for every man, woman, or child, God has a
your life that you may or may not agree with, yet a
t He has ordained through his providential control of
d, and that according to his sovereign will; a plan for
us, whether elect or non-elect, whether tall or short,
or, king or common.
the initial act of creation of the earth, God effectually
His own, each one in turn, at a particular moment in
hether you are a Christian believing in the saving
Christ, or not a Christian believing in whatever turns

11

THE NEW BIRTH ZYGOTE: *THE BABES*

"Yea; have ye never read, Out of the mouth of babes and suck-lings thou hast perfected praise?" (Matthew 21:16)

The first two chapters from the book of Genesis desig-nates God as a master Designer/Architect/Physicist, who constructed the entire blueprint of the universe; and also as an exceptionally genuine and efficient Chem-ist/Biologist/Geneticist, who invented and assembled strings of chemical components known as DNA. Many well-known scien-tists have ruled out the possibility that this remarkable, mys-terious entity could have appeared in this manner. However, the Bible emphatically states in Genesis 2:7 that: *And the LORD God formed man of the dust of the ground, and breathed into his nostrils the breath of life.* Could "dust" refer to the basic constituents of several chemicals and molecules, includ-ing nucleic acids consisting of nitrogenous bases, phosphates, and sugars—the building blocks of DNA?

The living soul, empowered and upheld by the Holy Spirit, is the result of the union between the body and spirit. It is an entity resulting from the joining of the unique sequence of nu-cleotide bases in the DNA molecules, created by the Father, and the informational program by which it functions, designed by the Son.

The most fundamental reason why each of us has such a distinctive personality is because at the moment of conception, God fashioned each one of us to be unlike any other person. The living soul does nothing apart from the unique sequence of the informational content embedded in the DNA structure. Humans and monkeys may have similar genetic sequences, but they have vastly different embedded information, vastly distinguishing humans from monkeys. Evolutionists harp upon this genetic similarity and loudly proclaim the identity of our recent biological ancestors, ignoring the science that embedded information makes a profound difference.

SPIRITUAL CONCEPTION

From corruption and mortality, to incorruption and immortality, seems to be the essence of our existence. Mortal man births, marries, produces offspring similar to his nature, and then dies. Children are a blessing to mortal man, but to Christian parents, birthing of spiritual children is impossible. God determines those who are to experience the new birth. The conception, the birth, and the life of Jesus are our guides for the conception, the birth, and the life of the spiritual child of God and may serve as our prototype.

FOUR BIOLOGICAL PHASES

Biologists have exposited four levels or phases of genetic activity. We have discussed the first two levels, the Molecular level where cancer and predestination genes reside, and the Cellular level where we are composed of 60 trillion cells divided into about 200 different types. Different cell types inhabit different organs: liver cells in the liver, kidney cells in the kidneys, and blood cells in the blood.

The remaining two levels of the biological phases are the Organismal and Populational levels. The organismal level includes all that goes on in the molecular and cellular levels; where we now start to observe several characteristics of the cell, such as the red color trait in the bloom of a flower. The red color observational trait is from a pigment in the cells that, when looked upon, gives the organism its red color. This extends to biological traits of organs—livers, kidneys, and hearts. Lastly, the occurrences of a trait within a species are observations at the Populational level, such as a large bed of red roses, analogous to a large population of people.

All living things go through this four-fold process that determines the reality we see in the world today: firstly, DNA

you on, a belief in Christ qualifies a person to receive a plan for both this life and the life to come.

Being effectually called is a calling that produces a spiritual conception in the body, i.e., *regeneration* → *molecular* → *zygote* → *babes,* an initial process that continues throughout the lifetime of a Christian where it acts primarily upon the molecular structure of the body, ultimately culminating in a physical zygote, a theological babe. It is a rebirth, a new beginning, a new order, that brings in a new life; a *continuing* transformation of the *old man* into a *new man,* nDNA to sDNA, cell by cell; until the process is completed and man is finally rebirthed, *born again,* or *born from above*; a never before seen *spiritual* entity in close organic relationship with God:

> From all eternity God purposed to have a family circle of His *very own,* not only created but *also generated* by His own life, incorporating His own seed, "sperma," "genes," or "heredity." Long ago, even before He made the world, God chose us to be *his very own* [in a genetic sense], through what Christ would do for us.[2]

Regeneration is both an *instantaneous act* and a *process.* It is the instantaneous creation of a new man of spiritual sDNA, and a process of continual, instantaneous creations of spiritual sDNA molecules, until the *fulness of Christ* (Ephesians 4:13) in you is complete. This progression does not imply that the work of the Spirit is incomplete, but that more natural cells are being regenerated to spiritual cells on a *continuing* basis throughout the lifetime of the Christian. Each regenerated cell *adds* to the stature of the body, hence the old nature is not improved, purified, or made whole; but changed, transformed, or metamorphosed. The Christian can neither increase nor decrease in the *quality* of substance, but will increase in *quantity* of substance. By this, regeneration is both an instantaneous act during *each* transformation, but a lengthy process acquiring *additional* lifetime transformations.

MOLECULAR

Natural and spiritual lives have their origin at the molecular level. In natural life, this is where the male sperm comes in contact with a female ovum, the joining of which produces a new and unique individual never before seen in existence.

Human beings consist of 46 chromosomes, the number necessary for a single individual member of the human species. Natural human life starts immediately *after* the act of fertilization. No longer is there a spermatozoon and an ovum, but a *human being.* In fertilization, a male sperm and a female ovum are needed. Both possess human life, since they both originated in separate, living, human beings. But they are not each whole human beings themselves, for a sperm can produce only sperm-type proteins; an ovum only ovum-type proteins. Neither acting alone can produce a complete human being.

ZYGOTE

Jesus had a natural biological mother of sinless ova, and a natural biological father, as all of us have had. Mary conceived by means of sinless, inseminated spermatozoon of King David by the Holy Spirit; my mother conceived by my father of sinful spermatozoon. Both mothers bore a child through natural birth. Mary bore a spiritual child—Jesus; my mother bore a natural child—me. Jesus was born sinless; I was born sinful.

Both mothers had a zygote, an amniotic sac, a placenta, an umbilical cord, and all the features of fetal cells that go to make up a pregnancy. Both mothers contributed nutrients to their respective embryos, but maternal blood *never* crossed the placenta to pollute the fledgling organisms; for blood in the zygote comes from the father, never from the mother! By this process, my natural zygote and the spiritual zygote of Jesus were protected; ere His precious blood would be contaminated!

I was conceived and born as an *old man* in sin, in my natural birth process, but now I am regenerated as a *new man* in true righteousness and holiness, in my spiritual birth:

And that ye put on the new man, which after God is created in righteousness and true holiness (Eph. 4:24).

Our spiritual conception is in this wise: *Jesus paid it all!* He purchased you and me with the value of His precious blood. Your whole spiritual being was sought-out, bought-out, and thus *conceived* with a price:

What? know ye not that your body is the temple of the Holy Ghost which is in you . . . and ye are not your

own? For ye are bought with a price: therefore glorify
God in your body, and in your spirit (1 Cor. 6:19–20).

Your conception is *the redemption price of the slain Lamb*
of God! Where did it happen?—On Mount Calvary. When?—
In 33 A.D. when He died. You now no longer belong to your
natural body although you remain associated with it and attached to it. Your body is the temple of the Holy Spirit who
resides in your spiritual body, not in your natural body. The
phrase, *not your own,* is a reflexive pronoun that agrees with
its noun, *ye,* before it and may be translated as, *and ye are not*
of yourself; "not of yourself" in that you no longer belong to
your natural body, but to your spiritual body.

This process is our spiritual conception. We are totally conceived by God in a totally new makeover of the basic molecules
of life, purchased by the infinite value of the precious blood of
Christ. Whether it is a natural conception or a spiritual conception, life begins at conception! Anyone purchased by the
blood of Christ is *conceptually reborn,* effective immediately,
alive eternally. This metamorphosed, spiritual entity—the zygote—the product of God acting upon the human cell is a miracle akin to the following:

• Of the Father's participation in the act of conception in
adjusting the molecular structures of our predestination genes,
analogous to the natural spermatozoon effectually calling out
those ova with the same specificity as that of the elect—as the
Father penetrates the Plasma Membrane of the cell to perform
a sovereign act of metamorphosis to reinstitute the "image of
God" into the genome; the initial step in the divine act of regeneration.

• Of the Son's participation of providing the spiritual
knowledge to the genetic code to lead a spirit-filled life, is akin
to the ebbing and flowing of the natural ovum, flowing in concert and synchronism with the actions of the spermatozoon to
reposition nucleotide sequences—as the Son orders the myriad
instructions to the ribosome apparatus to effect transcription,
translation, and expression of spiritual m-, t-, and r-RNA.

• Of the Holy Spirit's participation of providing the means
whereby both come together in conception and fertilization to
form the zygote—immortal, incorruptible, eternal—is akin to
the fusing and transformation of both spermatozoon and ovum

in producing the zygote, initiating the first signs and appearances of the *new life* of the *new man* in the *old man*.

BABES

Regenerated spiritual zygotes, as viewed in the natural biological realm, find their Biblical counterparts in the persons of newborn babes, who were just born and who have just begun their spiritual life:

As newborn babes, desire the sincere milk of the word, that ye may grow thereby (1 Pet. 2:2).

It is not uncommon in the Scriptures to compare Christians with little children and as newborn babes (Gk. *brephos*); the command now is to begin to grow in your spiritual walk by ingesting the *sincere milk* of the Word of God. The Apostle is not here calling us "baby Christians," but is telling us to earnestly yearn for God's Word, just as newborn babes cry for their mother's milk. This hunger for the Word should epitomize all Christians from the newly born to the most mature, for in it he loves the truth, is nourished by it, and will experience stunted growth without it.

Whether it is natural newborn or spiritual newborn babes, milk is the preferred mode of nourishment. That is where the analogy ends, for natural milk for natural babes is very different from sincere milk for spiritual babes. It is not philosophy that is needed at this stage, nor is it profound and difficult doctrines of the gospel; it is those elementary truths that can be comprehended by children by faith alone.

These are new spiritual babes who have just begun to live at the most basic level of the Christian life. The pursuit of Christ starts here. To know the will and the mind of Christ starts here. It is a compelling craving as evidenced by the Psalmist: *As the hart panteth after the water brooks, so panteth my soul after thee, O God* (Psalms 42:1).

Our spiritual process is now froth with injections of divine guidance. Every step, every utterance, every act committed in the flesh, is noted and recorded for future correction, for babes are not yet able to function independently of the Christ who redeemed them. We are admonished to go on in life for it is by slow degrees that he measures up to the full stature of Christ.

This is an excellent example of the grace of God as He plants *seed* in that part of the body that will ultimately respond to His calling, producing telltale marks of a Christian:

• The Elect of God
Elect according to the foreknowledge of God the Father, through sanctification of the Spirit (1 Pet. 1:2).

In Matthew 13, there is an excellent description of the spiritual selection and growth of an elect individual. In it, the author establishes the relationship of the biological growth of a seed with the spiritual growth in the new birth. The Apostle describes three different types of people and their problem with hearing; *the way-side hearers, the stony-ground hearers, the thorny-ground hearers,* and the fourth, those who can hear the call and thus respond to it—*the good-ground hearers.*

The parable is concerned, not with the sower of the seed, but with the type of seed and with the various soils in which the seed is sown. The Lord Jesus explained the different soils as representing various classes of those who hear the Word, four in number; and may be classified as *hard-hearted* (v. 19), *shallow-hearted* (v. 20–21), *half-hearted* (v. 22), *and whole-hearted* (v. 23). Three of the types, the way-side, the stony-ground, and the thorny-ground hearers, are those who, for various reasons, do not, nor can they, respond to the Word of God:

• The wayside hearers (the hard-hearted)
And when he sowed, some seeds fell by the way side, and the fowls came and devoured them up (v. 4).

• The stony-ground hearers (the shallow-hearted)
Some fell upon stony places, where they had not much earth: and forthwith they sprung up, because they had no deepness of earth (v. 5).

• The thorny-ground hearers (the half-hearted)
And some fell among thorns; and the thorns sprung up, and choked them (v. 7).

But those who do respond to the call of God are:

• The good-ground hearers (the whole-hearted)

But other fell into good ground, and brought forth fruit,
some an hundredfold, some sixtyfold, some thirtyfold (v.
8).

Verse 8 adds more clarity: *But other* [seed] *fell into good*
ground, and brought forth fruit. The Greek word for "good"
(kalos) has the meaning of morally valuable or virtuous ground,
and for "other" *(*Gk, *allos)* has the meaning of *different* seed;
different from the type of seed that fell by the wayside, upon
stony places, or the type that fell among thorns. Those seeds
produced no fruit but withered and died. The *other good seed*
produced an abundance of good fruit. The field is each person;
the sower is Jesus; the seed sowed is His spiritual sDNA, sown
in the natural nDNA flesh of those called to be saints:

He that soweth the good seed is the Son of man; The
field is the world; the good seed are the children of the
kingdom (vv. 37–38).

The "good-ground hearers" are the babes in Christ, those
who have been predestinated, called out, regenerated, passed
through the molecular phase of growth, been fertilized as a
zygote, and now are enabled to hear the Word of God and respond in faith, for we now possess:

• His Incorruptible Seed
Being born again, not of corruptible seed, but of incor-
ruptible, by the word of God, which liveth and abideth
forever (1 Pet. 1:23).

We bare an express image of Christ by means of our new
birth by conception of incorruptible seed. The Greek *spora*
(seed), as used here, and this only once in the Bible, does not
allude to a person's offspring; it looks back the other way, alluding to a person's *parentage*, to Christ. This seed has reference to bodily structures; nucleotides, genes, DNA and
chromosomes. Christ is our parent and His *incorruptible spora*
is in our spiritual sDNA.

When a person accepts Christ and is born again, he has a
new seed planted within him from God, and that seed is lodged
within every cell that is regenerated, within the very molecular structure of his being. This is the seed that is incorruptible.

Incorruptible seed means incorruptible life! This seed was never born, so it can never die. The life it brings was never born so it cannot die. In order for life to be eternal, it must also be incorruptible. Incorruptible seed implies eternal life.

Every seed must produce after its own kind. The seed, when it produces, and what it produces, is not anything different than what it is. The incorruptible seed of God in us does not reproduce anything but an incorruptible being; it does not produce a fallen soul and a corrupted body. This seed is incorruptible and the genes in this seed must be incorruptible, and they are. If we are to live forever, and we will, then we must be gened with genes eternal:

> Mom and Dad's genes won't do; we must be gened from God. . . . The first time we are born of temporal genes. If we live forever, we must be born of eternal genes. . . . God is saying here that the Eternal must gene you if you have eternal life. The Beginning must gene you. That which always was must gene you. . . . Eternal life cannot spring from corruptible seed, and nothing that one can do to the life born of corruptible seed can make it incorruptible and thereby eternal, nor can that which is done by corruptible seed nullify that which was wrought by incorruptible seed, and a life that sprang from incorruptible seed cannot be nullified by a life that springs from corruptible seed, which means the deeds of the natural man cannot stop the life of the spiritual man.[3]

His Word is the means of spiritual life and perseveres in it, till it brings us to eternal life; it remains eternally true, and abides in the hearts of the regenerate forever.

> A skeptic once told Gaylord Kambarami, the General Secretary of the Bible Society of Zimbabwe, "If you give me that New Testament I will roll the pages and use them to make cigarettes!" Gaylord replied, "I understand that, but at least promise to read the page of the New Testament before you smoke it." When the man agreed, Gaylord gave him the New Testament and that was the last he saw of him for 15 years.
>
> Then, while Gaylord was attending a Methodist convention in Zimbabwe, the speaker on the platform suddenly spotted him, pointed him out to the audience and

said, "This man doesn't remember me, but 15 years ago he tried to sell me a New Testament. When I refused to buy it he gave it to me, even though I told him I would use the pages to roll cigarettes. I smoked Matthew and I smoked Mark and I smoked Luke. But when I got to John 3:16, I couldn't smoke anymore. My life was changed from that moment!" That man is now a full- time evangelist, preaching the Word he once smoked! [4]

• Our Sins are Forgiven

I write unto you, little children, because your sins are forgiven you for his name's sake (1 John 2:12).

In John's day, he wrote to all classes of believers. Here he addresses those in the congregation, the *little children* or *babes* who were the very young in their spiritual life. They are the spiritual infants, whether they be physically full grown or just born. He addresses those who are probably under the age of twenty years since they were younger than the "young men," a word usually applied to those who were in the vigor of life; from about the age of twenty up to forty years, that needed exhortation that their future conduct should exemplify their position in Christ.

As each of us begins the Christian life, we know that our joy is found in forgiveness, and we know right well that God's forgiveness does not come by degrees. Even the youngest babes in Christ are completely forgiven; none are left behind; they can never be "more" forgiven. Forgiveness is an eternal gift.

This is why John wrote to them, that *Your sins are forgiven you*, that they should grow to maturity to assume their position in life under the headship of Christ. We should shower our little ones in Christ with the same message; to stress the fact that God has mercy on us and loves us; thereby we know that God is benevolent as not to impute our sins, and to rely solely on the name of Christ in whatever we say or do.

• We cannot Commit Sin

Whosoever is born of God doth not commit sin; for his seed remaineth in him: and he cannot sin, because he is born of God (1 John 3:9).

We come now to one of the most difficult verses in Scripture. We have repeatedly maintained the truth, as we understand it to be, that a Christian *cannot* sin in his *spiritual* nature, but may sin *non-habitually* in his *natural* nature only.

Many expositors indicate that a born-again Christian may sin in his body, in both natures, but to do so in a non-habitual manner. They maintain that in this line of reasoning, a Christian cannot distinguish between the two natures in the one body and, therefore, when he performs acts of sin, the whole person, including both natures, may be responsible.

But we answer that a Holy Spirit regenerated spiritual nature cannot sin, as our passage plainly indicates. What don't you understand about *doth not* commit sin? Is the poor Holy Spirit unable to secure the new man from sinning? No! A Christian who can sin in his spiritual nature is no Christian at all, for His seed *would not remain*! But John says that *his seed remaineth in him!*

- *But ye are washed* (1 Cor. 6:11a).
Regenerated, reconciled, set holy in your sDNA flesh.

- *But ye are sanctified* (v. 11b).
Set apart from the rest of the world in your nDNA flesh, for the inward cleansing and purifying of sin in the soul in regeneration, and set holy in your sDNA flesh.

- *But ye are justified* (v. 11c).
Made sinless in sDNA flesh by the power of the Spirit.

- *Ye are the salt of the earth* (Matt. 5:13).
Salt cleanses and purifies.

- *Ye are the light of the world* (v.14).
Light opens the way for men to see the beauty of Christ.

CHAPTER THOUGHTS

It is admitted that our new birth conception is shrouded in mystery, as is the conception of our Lord Jesus. His conception may elude us, but his birth process is exactly as our own.

He once was a zygote, a blastocyst, an embryo, and a fetus, and finally a Son at birth. His conception was supernatural but his birth process was natural. Our conception is likewise supernatural and our birth process is natural, mimicking His birth process and culminating as a son of God. In between our conception and new birth, between our regeneration and glorification, lies a whole new world of living.

Our conception is obviously not identical to that in the natural realm, but I envision it to be similar in function. Who can explain adequately the physical and spiritual laws governing the birth process, the outcome of which is a living human being? What about of all those happenings going on, of all those little things we come to affectionately call gametes, ova and spermatozoa, and spermato- and oo-genesis; just where does all this come from—where does it all end?

Every mouth is stopped when we attempt to explain how the tiny sperm knows enough to wiggle its tail (the flagellum) to swim uphill in the fallopian tube to seek out this very inviting prey, the ovum, to sink its teeth into the "egg" for a nutritious meal, only to discover that it has been outwitted and thus instrumental in causing fertilization. The race is thousands of miles long with respect to the size of one sperm and the contestants number in the millions, dwarfing the largest and longest race in recorded history. What impels them to such behavior?

The very best we can do, then, is to formulate a spiritual conception that can be sustained, which is both biblically sound and scientifically feasible. Many Scriptural passages that have been traditionally referred to as belonging to accepted interpretation have been expanded in its basic meaning to incorporate a possible, deeper meaning. In this wise, the conception of a Christian has been referred to as a three-step process over against the two-step process of natural conception in the joining of spermatozoon and ovum.

This three-step process of conception consists of the three Persons of the Trinity participating jointly, having equal inputs, without which one missing would cause a miscarriage and void the whole. The Son purchased our redemption, and thus our conception, the moment He redeemed us on the cross at Calvary and bore our sins in His person. As with the sperm seeking out the ovum to form a new life, like a hound dog after its prey, so also the Trinity seeks out and impregnates His elect with the life of God.

Cognizance of what is happening to us during Phase 1 of our birth and development process is not found in our memory.

But don't fret; you were still a child of God, regenerated to newness of life by the Holy Spirit. You are no more a child of God at Phase 4 in our diagram than you are at Phase 1.

Incorruptible seed! What can we say about incorruptible seed? Every Christian is blessed with this amazing seed. It is the seed of God. It sustains no relationship to natural seed, indeed it is at war with the natural seed. Incorruptible seed cannot, nor will not, coexist peaceably with natural seed. One will war against the other, but the outcome is forever settled.

The zygote phase lasts but a short time, but long enough for the new man to be able to express some initial thoughts as to why he now feels the way he does; why there seems to be more joy and enthusiasm towards the things of God; why he acts different towards his old friends; and why it seems to hurt a little when desirous to engage in his old sins once again.

TIDBITS OF PART III

- The Trinity → One God, Three Persons; coequal:
  - The Father → The "I" Person;
    - The Energizer.
  - The Son → The "You" Person;
    - The Informer.
  - The Holy Spirit → The "He" Person;
    - The Regenerator.
- O.T. YHWH = N.T. Theos; God is God:
  - Eloah (God), Elohim (Gods), Elohenu (Our Gods).
  - The Shema; The LORD our God is One LORD.
  - Ehad & Yahid; "one" & "one"; both not the same!
- Jesus: Son of God and Son of Man; deity and humanity:
  - In Adam's genome.
  - Jesus' body → sDNA.
  - His biological father—King David.
  - The right to sit on the Davidic throne.
  - King of Kings and Great High Priest forever.
  - One sNature, one dNature (d for Deity).
- The Holy Spirit:
  - Overshadowed Mary with sDNA of David.
  - Overshadowed us with sDNA of Christ.
- A Christian → Reborn, regenerated:
  - A new species.
  - In Adam's genome.

- o One sBody;
  - ♦ With nDNA and sDNA flesh.
- o Two natures;
  - ♦ nNature, sNature–a new creature.
- o Transformed;
  - ♦ By metamorphosis.
- o sDNA flesh;
  - ♦ Witnesses to nDNA flesh.
- o Non-Elect have nDNA flesh;
  - ♦ One nBody, one nNature.
- Becoming Christian—Suspend Second Law of Thermo!
  - o Jesus;
    - ♦ Our prototype!
  - o We sin in our nDNA;
    - ♦ We *cannot* sin in our sDNA!
  - o Our redemption;
    - ♦ The purchase price of our conception.
  - o Whole hearted, good ground hearers;
    - ♦ Ye are washed.
    - ♦ Ye are justified.
    - ♦ Ye are sanctified.
    - ♦ Ye are the salt of the earth.
    - ♦ Ye are the light of the world.
- War!
  - o The SEED vs. the seed!
- Phase 1: *Regeneration → Molecular → Zygote → Babes:*
  - o Regeneration → Sinless antigen/pathogen complex;
    - ♦ DNA attacked.
    - ♦ Metamorphosed → sDNA.
    - ♦ By analogy of the simple virus.
  - o Molecular: The Seed of God;
    - ♦ Incorruptible! Immortal!
  - o Zygote: A fertilized ovum;
    - ♦ 2 cells, a human being!
  - o Babes: The Elect of God;
    - ♦ The good-ground hearers.
    - ♦ His incorruptible seed.
    - ♦ Our sins are forgiven.
    - ♦ We cannot commit sin.

PART IV

The Battle For The Body

12

LOVE'S LONG JOURNEY

"God is love" (1 John 4:8)

Specific to our topic is what role the individual members of the Trinity play in the work or plan of God in the creation and spiritual growth of individuals. Creation is the direct outcome of the love of God the Father for without His Love, there would be no creation; and out of that momentous creation came man, endowed with the image of God. As the Psalmist was perplexed with the question of why creation of man at all, considering the vastness and orderliness of the universe, he proclaimed:

> *When I consider thy heavens, the work of thy fingers, the moon and the stars, which thou hast ordained; What is man, that thou art mindful of him? and the son of man, that thou visited him? For thou hast made him a little lower than the angels, and hast crowned him with glory and honour* (Psa. 8:3–5).

Why is it that God is so *mindful* of man, that He should create the whole of the universe with all its hosts of stars and galaxies in order to house him; make dirt and grow food in order to feed him; instill in him a spirit in order to commune with him; and create a woman for him in order to make more of what He originally made? What are His plans, His motives, His desires? What is the overwhelming driving force that sustains these ambitious plans?

Perhaps it was that when the psalmist David looked up into the starry sky on a clear, cool night several thousand years ago, and saw the myriad of stars shining brightly up in the heavens, that his eyes came in direct contact with the awesomeness of his God. Thoughts of grandeur raced through his mind—what a magnificent God! *O LORD, our Lord, how excellent is thy name in all the earth!* (Psalms 8:1). How majestic are thy works, O LORD!

CREATION AND THE FALL

CREATION

Perhaps David did not know at that time that the Earth was 7,926 miles in diameter at the equator, 24,901 miles in circumference, or an average 93 million miles distant from the Sun. Perhaps he did not know that Jupiter is 11 times larger than the Earth and the Sun is 10 times larger than Jupiter. But David knew his God. He expected great things from God, and attempted great things for God throughout his life. As he gazed up and listened to the speech of the universe, to the speech of those things without life, as they cried out to him, he declared without hesitation:

> *The heavens declare the glory of God; and the firmament showeth his handiwork. Day unto day uttereth speech, and night unto night showeth knowledge. There is no speech nor language, where their voice is not heard* (Psa. 19:1–3).

As we end the day at nighttime, gazing up into the starry heavens pondering how all this can be, we are also faced with the wonder about the value that God has placed on man. The glory of God is magnified in His creation and the exalted position of man is the capstone of the entire universe. Wonder of wonders is the fact that the Lord God can hold the Milky Way in one hand and on the other, take infinite interest in you and me. Is it not unreasonable then, that when the time was right, when the moment was at hand, when the Son of God could wait no longer to *see his seed* (Isaiah 53:10), the Father then called into existence the entire universe: *In the beginning, God created the heaven and the earth* (Genesis 1:1).

It took God six whole days to finally arrive at the point where He would make man, and that after all of the preliminary creations have been completed in order to sustain his life. The Genesis account of the creation of man is well known to all:

And the LORD God [Elohim] *formed man of the dust of
the ground, and breathed into his nostrils the breath of
life; and man became a living soul* (Gen. 2:7).

It was Elohim who was dominant in the act of creation of
Adam, hence the Trinity is implied. In addition, it is God the
Father who is the predominant directive force acting here individually. He *formed man of the dust of the ground,* clearly
relating to flesh and bones we readily see with our eyes; but on
the molecular level, we see DNA, proteins, and cells, working
in synchronism to produce the first man. This *forming* of Adam is more glorious then that of the *creation* of the heavens
and the earth.

THE FALL

Many scholars believe that the Biblical book of Job was
written around the same time (could it have been Job?), or relatively near when Moses wrote the book of Genesis. Some even
go further and indicate that Job was written *before* Genesis!
Whichever is correct, it is interesting to note that some passages in Job speak directly to the introduction of sin into the
world that no other earlier book addresses.

In Genesis 2:9, Adam feared God and eschewed evil, and
early-on obeyed the commandment not to eat of *the tree of
knowledge of good and evil* lest he should surely die. Job
speaks of the Tree but in other scenes and in different episodes.
Here in Job, God and Satan take center stage, with Job acting
in place of the first man, Adam.

*And the LORD said unto Satan, Hast thou considered
my servant Job, that there is none like him in the earth,
a perfect and an upright man, one that feareth God,
and escheweth evil?* (Job 1:8).

He had *seven sons and three daughters* (v. 2), and had
many possessions: *seven thousand sheep, and three thousand
camels, and five hundred yoke of oxen, and five hundred she
asses.* Job *was the greatest of all the men of the east* (v. 3). God
has given to both Adam and Job blessings of family and material wealth beyond measure, with instructions to care for and
to subdue the earth.

Then one day, Satan came and stood before the Lord and Satan answered the Lord's question in verse 8, saying:

Doth Job fear God for nought? Hast not thou made an hedge about him, and about his house, and about all that he hath on every side? thou hast blessed the work of his hands, and his substance is increased in the land. But put forth thine hand now, and touch all that he hath, and he will curse thee to thy face. (vv. 9–11).

It is not God who tempts His people to sin. God "hedges" us in so that we are protected from committing evil acts. God is a God of Love, not of evil. Satan is clearly seen as the tempter and the author of sin—the father of all lies! Does this episode remind you of the passages in Genesis where Satan, as the serpent, casts a temptation before Adam?

God then gave Satan *permission* to do to Job as he saw fit, whatever pleased him to do. Anything was permissible, short of harming Job himself. Satan then proceeded to destroy Job's seven sons and all his possessions, but still Job did not blame God for his current condition.

Satan was not very pleased with the outcome of all the misfortune that befell Job, as Satan desired that Job would blame God to His face and thus loose his righteous standing before Him. Satan then proposed a second assault upon Job's character, saying:

But put forth thine hand now, and touch his bone and his flesh, and he will curse thee to thy face" (Job 2:5).

So Satan was again given permission to assault Job, this time upon his body, as he *smote Job with sore boils from the sole of his foot unto his crown* (v. 7). Satan did so, and his entire body from head to toe was infected. The very same scene and actual conversation could be said about Adam when he sinned in the Garden. God pronounced the sentence of death on Adam and all his descendants, but it was Satan who was responsible for sin to enter into Adam's DNA structure. God permitted it; Satan seized the opportunity to do it.

As Satan *touched his bone and his flesh* (v. 5), and *smote Job with sore boils from the sole of his foot unto his crown* (v. 7), so also was Adam's bodily structures affected with sin, from the sole of his foot unto the crown of his head. God allowed Satan to attack Adam in his genome as Job was attacked in his body. Adam's nDNA became infected. His predestination genes degenerated into pseudo genes, relegated to the realm of junk jDNA. Original sequences were realigned, spiritual information lost, the Second Law of Thermodynamics made effective, and the life of God in the Holy Spirit was as seen exiting from above the Holy Temple, exemplified by the scene in Ezekiel where the glory of God was seen by the nation of Israel hovering over the temple, then slowly seen ascending upward from the city; over the mountain tops, and then taken up out of view and out of fellowship with the whole nation:

> *Then did the cherubim lift up their wings, and the wheels beside them; and the glory of the God of Israel was over them above. And the glory of the LORD went up from the midst of the city, and stood upon the mountain which is on the east side of the city. . . . So the vision that I had seen went up from me* (Ezek. 11:22–24).

God will someday come again, the same way as He left, but this time He will not simply abide within the cloud hanging over the temple, but He Himself will reside within us for we are the Holy Temple of God. He will restore what is rightfully His. He will not remain passive indefinitely. Both Adam and Job were restored to a greater position than they originally held. They were blessed beyond measure. Both were essentially regenerated to a higher life. Regeneration ushers back in that which was lost. Adam's predestination genes were restored, void of all vestiges of degenerated pseudo genes, and became the progenitor of both the natural and the spiritual race. This scenario is repeated over and over in the lives of His spiritual children. The life of Job, his fall from grace and his subsequent rise to a higher position, is the gospel at the molecular level, where the fall is in his natural nDNA, and his rise is in his spiritual sDNA:

> *So the Lord blessed the latter end of Job more than his beginning; for he had fourteen thousand sheep, and six thousand camels, and a thousand yoke of oxen and a thousand she-asses. He had also seven sons and three daughters* (Job 42:12–13).

Will not God bless you more abundantly above all that you can imagine, simply because He elected you to be *in Christ!*

SIN AND THE IMAGE

SIN

It is reasonable then to assume that God permitted Satan to use the actions of mutations to cause detrimental changes in Adam's natural nDNA, thereby losing his potential for immortality. The introduction of sin knocked out nucleotide sequences along Adam's nDNA helical structure, modifying the amino acid code for immortality sustaining proteins. Erased also is the critical information associated with the embedded nucleotide code that sustains the ability of the body to continue in that state.

Biological mutations are changes in cellular DNA, permanently altering the genetic code for amino acid sequencing in proteins. Figure 12.1[1] illustrates how nucleotide bases out of sequence can cause disastrous effects to an organism.

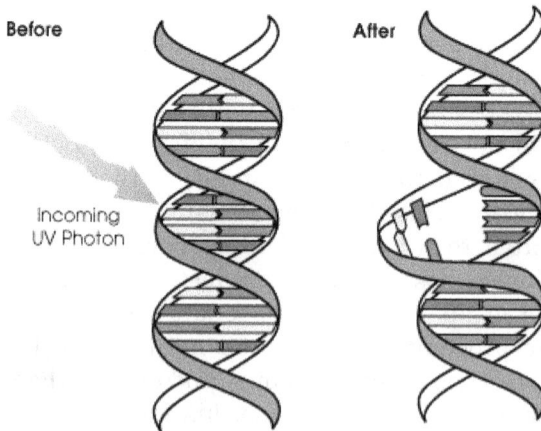

Before   After

Incoming
UV Photon

Figure 12.1: Nucleotide Mutations

Human nature stands or falls on the proper sequence of amino acid bases and to the information assigned to that sequence. The aberration in Adam's *very good* moral nature, a nature of such high and lofty magnitude, was instrumental in its immediate decline and decay to a natural, moral nature.

Adam's body was created sinless and, therefore, potentially immortal, and pronounced *very good,* but he was not made holy as the Father is Holy: *Be ye holy; for I am holy* (1 Peter 1:16). This verse is in the imperative mood that acts as a two-sided coin. On one side, it implies the command that this is what the Father sincerely wants His children to be—holy as He is Holy. But on the other side of the coin, there is another nuance of the Greek imperative, for the Father says that: "But I will not give you the *power* to be as holy as I am Holy," evidenced by the fact that no one has ever seen a man as holy as the Father (except, of course, the Lord Jesus Christ).

Holy in this sense applies to deity, which Adam was not. The new sequence now knows nothing of the morality of belief in God as He originally instituted in the genome, and all communication with Him has now been lost. Man in his fallen nDNA state is in a helpless condition, unable to assist himself in obtaining that which was lost.

Adam sinned and lost his potential immortality, became mortal, and thus all men through him sinned: *For all have sinned, and come short of the glory of God* (Romans 3:23). This action constituted a drastic change in Adam's DNA structure since all offspring come from the DNA content in both parent's germ line cells. His original nDNA was transformed through the actions of non-beneficial mutations which he then passed on to his firstborn son, Cain, and then passed down through the annals of history to include all men (Christ excepted).

Something sinister has now been added to Adam's mutated nDNA: *sin!* Sin is responsible for a vast array of misery, grief, and death that was not found previously but now runs rampant. It is the precursor of all diseases and manifests itself in mutations. It is a law in nature, a curse that is seen working in our cells:

> *But I see another law in my members, warring against the law of my mind, and bringing me into captivity to the law of sin which is in my members* (Rom. 7:23).

Man is now captive to sin and can do no other but sin. The disastrous effects and the relentless onslaught of entropy of the Second Law of Thermodynamics are now clearly seen in the manifestation of sin. At critical times in life, at the command of certain proteins summoned into action by God's initial plan for your life, non-beneficial mutations are seen to be the main cause of diseases. Most cancers are caused by mutagens, environmental agents such as ultra-violet light and chemical carcinogens that alter the nucleotide structure of the DNA molecule, hence the cancerous *Oncogenes*. This invasion happens at the Molecular Level and is illustrative of all mutagen invasions of genetic structures:

> Whole banks of genes have been discovered that control the cell's replicative machinery. Certain genes, *when activated*, cause cell division to occur. In cancer, these are the "Oncogenes," and there appears to be about 60-70 different ones in the human genome. The Oncogenes are held in check by the "tumor suppressor genes," which shut down the replicative process. It has been shown that most cancers arise through a multistep process. . . .multiple mutations may cause continuous activation of some Oncogenes and may inactivate some of the tumor suppressor genes. There is no doubt that the key word in cancer causation is "mutations," which lead to uncontrolled cellular growth. . . . In fact, it is becoming increasingly clear that without these mechanisms, we would all die of cancer in childhood.[2]

Non-beneficial mutations are *indirect* acts of God in that He uses *natural* occurrences to "turn on" the cancerous Oncogenes. Natural mutagens effect changes in increased entropy and decreased information in the human genome as determined by the Second Law. The sin mutagen has infected every human being born of woman (Christ excepted), and cannot be reversed in those who reject the gracious offer of salvation in Jesus. Most cancers originate in a single cell, dividing to produce daughter cells that are also cancerous. Once cellular growth has become malignant, they can invade normal tissue cells, metastasize, and migrate to other parts of the body.

However, on relatively rare occasions, *beneficial* mutations can and do occur. These types of mutations are the *direct* acts of God that result in the suspension or reversal of the Second Law, with the desired effect of decrease in entropy and increase in information content. As a result of this type of mutation, predestination genes are "turned on" and the effects of the curse of sin are dealt a mortal blow. Once cellular growth has experienced a new vitality with beneficial mutations, they can, as in the case of cancer, invade normal tissue cells, metastasize, and migrate to other parts of the body, reversing the degenerative effects of sickness and death.

THE IMAGE

Countless volumes have been written on what is meant by the statement that man is created in the image and likeness of God. God is Spirit and not a material being. We do not look like God physically. What we do bear of His image is primarily an invisible, inner likeness, a likeness not unlike mind, personality, and emotions. We possess a capacity to love and to express feelings, to will, and a moral responsibility with a capacity for worship. God made us very much like Himself, but to a finite degree.

The image of God in man has always been a magnet to attract theologians eager to express their views. When I was studying for the doctoral degree in engineering many years ago, I vividly remember spending countless hours down in the "dungeon", a laboratory (so to speak) where I performed many experiments in the fledgling scientific field of Holography.

Holography is the art of taking pictures with a laser instead of a camera, Figure 12.2[3], and the procedure is essentially the same. To take a picture with a camera, point it at the image or object you desire to capture, and then click the shutters. Develop the film and out comes the picture of the object. Simple! Holography is the same but uses different components.

First, there is a laser emitting a source of light, essentially of one frequency (monochromatic), whereas the camera uses a light source emitting many frequencies (the Sun). Secondly, place the laser so that it emits a beam of light that passes through a beam splitter that will split the beam into two individual beams; one beam that will illuminate the object, the *illumination beam*, and the other beam, the *reference beam,* that will impinge directly upon the film. Turn off the overhead room lighting (eliminating the source of many frequencies) so that the lab is in total darkness, and then remove the film protective covering (the same as opening the shutters) so that the reference beam impinges upon the surface of the film.

My very first experiment was taking the picture of the King and Queen Chess pieces, a classical experiment at that time. The illumination beam impinges upon the king and queen, captures and records the total image in the *object beam* that is reflected. The object beam contains every aspect of the king and queen pieces, but is of no use unless it interacts with the reference beam at the face of the film.

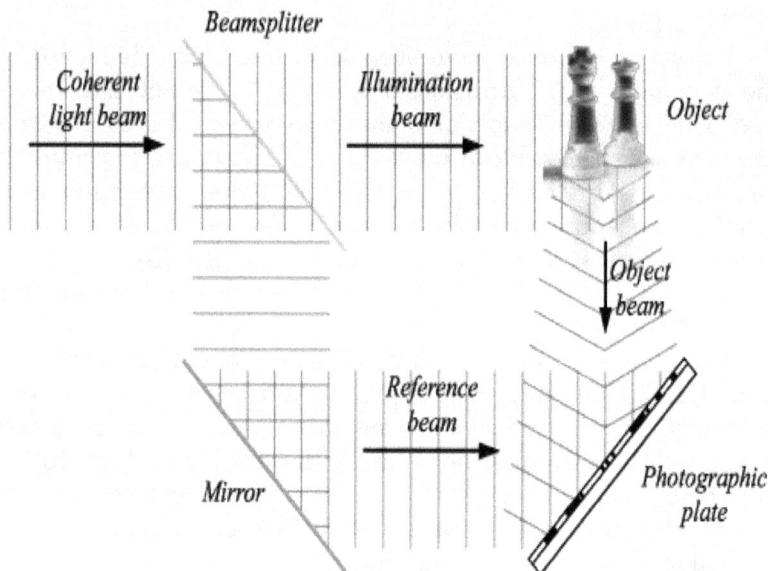

Figure 12.2: Holographic Recording

Develop the film and what do you see on the film? Nothing! Just wiggly lines, no king and no queen! What happened? Well, we're not done yet. Turn the laser back on and hold the developed film in front of the reference beam and now look at the film. Now what do you see? The king and queen!

You may at this point ask why I went through all this intricate and expensive way to take a simple picture of a king and queen. Why not just use a camera? Well, because a hologram is extremely pertinent to the new birth process that we have been talking about all along. For instance, take a pair of scissors and cut the film taken from the camera in half and retain only one piece. Now look at the half film piece, what do you see? You see half of the original image of the king and queen. Cutting the film in half from a camera is not a good thing to do as it modifies or destroys the original image.

Now do the same to the hologram film; cut it in half, throw away one piece, and retain the remaining piece. What do you see now? You see the *totality* of the king and queen! Don't be-lieve me? Spend a few hundred dollars for a laser and try the experiment yourself. Let's go further, keep looking at the re-maining piece of film as you continue to cut each piece in half, until only the smallest minutia of film remains. Now, what do you see? You still see the king and queen! Not only that, but slowly rotate the film and you will begin to see around and in back of the king and queen. As you rotate, you will begin to see the Pawn chess piece that I placed behind the king and queen that I didn't tell you about. Try that with the camera!

A hologram is truly a 3-dimensional image that captures 100% of the information about that object, front, back, and sideways, as long as the illumination beam impinges upon the total volume of space surrounding the object and deposits that information onto the film. Theoretically, the image of the king and queen (and the pawn) is captured down, and on to, the *very last molecule of film*. Allow me to repeat; the total image of the king, queen, and pawn is on every molecule of film! Loose the film but have one molecule cling to your clothes and you still have a picture of the image that you desired to take, clinging to your clothes!

> A small portion of a hologram's surface contains enough information to reconstruct the entire original scene. . . . This is possible because during holographic recording, each point on the hologram's surface is affected by light waves reflected from all points in the scene, rather than from just one point. . . . To demonstrate this concept, you could cut out and look at a small section of a recorded hol-ogram; from the same distance you see less than before, but you can still see the entire scene.[4]

The human body is a hologram. Look at one of your cells (if indeed you could) and you will see the image of God embedded in every sDNA molecule! Rotate the molecule and you will still see the image of God; it cannot be erased! Loose one cell and you still have billions more with the same image. Loose all the cells except one and you will still see the image of God. Illumi-nated by the laser Light of the Son of God, the image of God is forever retained in every molecule: *That was the true Light, which lighteth every man that cometh into the world* (John 1:9).

We have stated many times in the Genesis passages that man was created in the image of God. The Hebrew word for image is *tselem* which is a good translation but it really does not convey a precise meaning of the phrase, "created in the image of God." The primary meaning of tselem is that of a *shade* or a *shadow* of an object.

The "image of God" in our sDNA is but a *shadow* of who He really is, a representative figure of the Thrice Holy God. When God acts, we act in his shadow. If we may liken God to a mighty Oak tree with all its splendor and beauty, swaying majestically in the breeze, then we are but the *shadow* of it. Jesus Christ represents God to an infinite degree; we represent God to a finite degree. Might this *tselem* be the object beam illuminated by the laser Light of God?

Adam lost this image due to the sin he committed. The effects of sin upon humanity are physically analogous to the film from the camera cut in half several times, illuminated only with the light of the sun. God immediately permitted the sin pathogen to invade his body, causing a mutation in his DNA in every cell. Annihilation of the image to some degree was assured. Whether that annihilation was partial or complete, the fact remains that man lost, among other things, his ability to communicate with his creator.

I strongly believe that the image of God no longer resides in the genetic structure of the human body of unbelievers who know not God, whose body is the temple of Satan. I believe that God resides in the body of believers, never to be erased, where He dispenses abundant acts of grace and blessings, as stated in both the O.T. and with its corollary in the N.T.:

The LORD [the Father] *bless thee, and keep thee: The LORD* [the Son] *make his face shine upon thee, and be gracious unto thee: The LORD* [the Holy Spirit] *lift up his countenance upon thee, and give thee peace* (Num. 6:24–26).

The grace of the Lord Jesus Christ, and the love of God [the Father], *and the communion of the Holy Ghost, be with you all. Amen* (2 Cor. 13:14).

Be with you all!—with His image in every molecule in your body, in your soul, and in your spirit; in every aspect of your life—that is, if you are Christian; for hath not Jesus expanded on this and said to us:

Lo, I am with you always, even unto the end of the world (Matt. 28:20).

How else can Jesus *be with you all* and *with you always* if He is not embedded in your molecular structure?

If you are not Christian, then there is another scenario that can be said of you, for *Ye are of your father the devil, and the lusts of your father ye will do* (John 8:44). If Christ be not in you, then, sad to say, the devil is in you and you will do his bidding. It may be said that the image of Satan, your father, is embedded in every molecule of your body!—a very sad commentary.

CHAPTER THOUGHTS

Why is it that God is so mindful of you and me? Why should we deserve such an exalted position in the affairs of an infinite and loving God? Why create at all?

Is it because He is lonesome and seemingly bored with His life, extending from eternity past and reaching up to the date of the creation of the world? And now, fast forward to the climax of recorded history in the unforeseeable future, where we will be with Him for all eternity future, He will lavish His infinite love upon us. Why?

There is only one answer to these questions, attested to in Scripture and confirmed by millions who have eyes to see and ears to hear: God created because He loves! And those He loves He predestined to eternal life with Him. He did so way back when there was no time, no space, no earth, no nothing.

He could have created man as He did the angels who never sinned, who beheld His glory day and night, but He didn't. Some of them fell, who looked upon God, and sinned! Lucifer comes to mind. How could this ever be? How could angels in the presence of the Lord ever want to sin? Did they see something better in sin that they coveted, that they saw not in God? Was God, to them, inferior?

I have news for you—those angels who sinned before the very eyes of God are no longer in heaven. God demands love, not rote obedience. Then again, why is it that there are some who will and there are some who won't? What is materially wrong; what is missing in the makeup of man that demands him to act the way he does?

Is it because God knows right well that man, left to himself can't make the proper choice to choose God or not to choose Him. Is it because all mankind can do no other than *not* to choose Him?

From eternity past, God desires man to be in fellowship with Him, to be made with some degree of His image that this fellowship may be a reality. Man left to himself cannot do it. Even in the future Millennial Kingdom, when Christ physically sits on the throne to rule the world, He will show to everyone, even when sinful men will physically see Him, the God of creation, that they will reject Him. Even at the tender age of 1,000 years, man left to himself cannot choose God. In sovereignty, He predestined those whom He chose, instilled within them the image of His presence, and left the rest of mankind to themselves, all who willfully reject His love.

13

Biological Armageddon

*"From whence come wars and fightings among you? come
they not hence, even of your lusts that war
in your members?"* (James 4:1)

Our passage clearly indicates that there is a war in the
"members" of our body, two rivals, one energized by the
"lusts of the flesh," identified as natural nDNA flesh,
and the other energized by the "fruit of the Spirit," identified
as spiritual sDNA flesh and that energized by divine decree.

The elect Christian is in unceasing conflicts with three
major foes, namely: "the world, the flesh, and the devil." The
combats with the world and the devil are waged from without,
but the strife opposing the flesh operates from within. The old
flesh is constantly being destroyed; tissue is consistently being
transformed; for nDNA flesh is still a living part of every believer, and will continue in possession of its fallen nature.
sDNA flesh cannot, nor will not, live peaceably with nDNA
flesh; but will continue in conflict until the final redemption of
the body, at which time the spiritual flesh shall rise victorious.

The unregenerate, the non-elect, is never in need of a battle within the body, as there is no occasion for the flesh to rise
up against its own mind to force one to conform to the dictates
of another. No war is necessary between members that agree
with each other; no battle need be fought between allies that
wallow in the same sinful lusts and desires. nDNA flesh is capable and very happy to continue to function in its sinful
nature; receiving evil thoughts and emotions via neurotransmitters, while transmitting to the brain evil lusts and desires.
Why go to war when the flesh and the mind are satisfied with
the status quo; no need to go to war when there is no spiritual
entity abiding.

THE BATTLE ARRAY

Every person who is born of the Spirit passes through the fire. Each one of us passes through our own Battle of Armageddon, whether it be fought in the Valley of Jezreel, in the land of Israel, prior to the institution of the Millennial Kingdom of Christ at the end of the age, or whether it be fought within the human body of those who are regenerated. It is the battle to end all battles.

THE LOCATION

Battlegrounds for this final battle have both similarities and dissimilarities. While the one battle is to be fought in the future, the other is started and fought immediately when one is regenerated. One battle is fought in the land of Israel while the other is fought within the body of a Christian. In the first, the armies swarm upon Israel from the North Country, hoping for a swift victory over the little nation with unwalled villages:

> *And thou shalt say, I will go up to the land of unwalled villages; I will go to them that are at rest, that dwell safely, all of them dwelling without walls, and having neither bars nor gates, To take a spoil, and to take a prey* (Ezek. 38:11–12).

In the second, the armies of nDNA flesh swarm down upon the meager establishment of sDNA flesh, hoping for a swift victory over its unwalled cells, *to take a spoil, to take a prey.*

The battle will be fierce. The troops of the nations attacking from the north countries are commanded by *Gog, the land of Magog, the chief prince of Meshech and Tubal* (Ezekiel 38:2), identified today as modern Russia. They will swoop down upon Israel to take a spoil. They will enter the city of Jerusalem, unfortified, with unwalled villages, having neither bars nor gates for protection. The defenders are unprepared for war. Victory seems inevitable for the invaders.

Whether it is the fledgling nation of Israel or the sDNA body of a Christian, God has plans of His own. God will fight forever to preserve their identity, for *the LORD hath chosen thee to be a peculiar people unto himself, above all the nations that are upon the earth* (Deuteronomy 14:2). He will utterly destroy every invading enemy that attacks; victory is assured, for they are His *chosen people, a peculiar people.*

THE COMBATANTS

The main thought here is that God also addresses Christians as being a chosen generation, a peculiar people: *But ye are a chosen generation, a royal priesthood, an holy nation, a peculiar people* (1 Peter 2:9). Those who seek to destroy Christians will inevitably come face to face with the holy and righteous God. Whether it is on the battlefield at home, on some foreign soil, or fought within the body of each one of us in every nation of the world, victory is assured because our God is mightier than their god.

God will fight the battles of the Christian just as he fights for the survival of Israel, for without Him the battle is lost. As the hordes from the north countries are mightier in number than Israel on the battlefield, so also are the hordes of nDNA mightier in number than sDNA residing within.

Both battles have armies in combat readiness; both armies have battle armor for protection; and both armies are well trained and fortified. Both armies have troops that will fight to the bitter end in order to maintain their supremacy over the other. Both armies are fighting for one way of life or another way of life; between the way of life of those who belong to Satan, or of the way of life of those who belong to Christ.

When salvation in Jesus is come, war is immediately declared! sDNA forces are conscripted and mobilized in the newly ordered military draft; weapons are stockpiled and set in array; generals are assigned to their command posts; and captains of the armies are positioned in the field, ready to receive orders to go on the attack. The flesh now faces a new battle, a spiritual battle, and arrays itself against sinful nDNA armies that are numbered as the sands of the sea. Extremely well fortified with evil thoughts and lusts, nDNA in the flesh array themselves against the sDNA intruders that arise in the new birth and now find themselves in a battle for their lives; a battle well described as the "mother of all battles", the beginnings of a spiritual warfare for the control of the body and the soul.

The battle rages within. The battle rages in your mind, in your feelings, in your emotions, and in your flesh. It is a battle between the forces of darkness and the forces of light. It may feel to you that you are being pulled one way or the other, from one pole to the other pole. The conflict inside you may seem at times almost unbearable, but be strong and not faint. Look to God's Word continually to see what spiritual warfare is all about; to understand what is happening inside each of us.

This conflict is between the true God of the universe, the God of freedom and life; and the false god of this world, the god of evolution and death. The battle within you is between the Word of God and the radical changes that are proposed in your flesh, which spills over into the mind with its thoughts and emotions—against the babblings of Satan with his lusts of the flesh. You are in the middle of this spiritual battle; have faith and trust in God for the victory:

> *Submit yourselves, therefore, to God. Resist the devil, and he will flee from you. Draw nigh to God, and he will draw nigh to you. Cleanse your hands, ye sinners; and purify your hearts, ye double-minded* (Jas. 4:7–8).

John Bradford, a preacher of yesteryear, characterizes the combatants of this battle now raging, a battle between the *natural man* and the *spiritual* man (1 Corinthians 2:14–15); between the *natural body* and the *spiritual body* (15:44); between the *outward man* and the *inward man* (2 Corinthians 4:16); a battle between the *old man* and the *new man* (Ephesians 4:22, 24):

> A man that is regenerate and "born of God," consisteth of two men (as a man may say), namely of "the old man," and of "the new man." "The old man" is like to a mighty giant, such a one as was Goliath; for his birth is now perfect. But "the new man" is like unto a little child, such a one as was David; for his birth is not perfect until the day of his general resurrection.
>
> "The old man" therefore is more stronger, lusty, and stirring than is "the new man," because the birth of "the new man" is but begun now, and "the old man" is perfectly born. And as "the old man" is more stirring, lusty, and stronger than "the new man;" so is the nature of him contrary to the nature of "the new man," as being earthly and corrupt with Satan's seed; the nature of "the new man" being heavenly, and blessed with the celestial seed of God. So that one man, inasmuch as he is corrupt with the seed of the serpent, is an "old man;" and inasmuch as he is

blessed with the seed of God from above, he is a "new man." Inasmuch as he is an "old man," he is a sinner and an enemy to God; so, inasmuch as he is regenerate, he is righteous and holy and a friend to God, so that he cannot sin as the seed of the serpent, wherewith he is corrupt even from his conception, inclineth him, yea, enforceth him to sin, and nothing else but to sin.[1]

The Christian body, the one that you see when looking in a mirror, consists not only of one singular, monolithic person, but of an unholy assortment of diverse "persons"; not one but a body of two people, one person consisting of the "old man" and the other person consisting of the "new man." The whole of the body is the Christian body, no longer to be called a Jew or Greek: *There is neither Jew nor Greek, there is neither bond nor free, there is neither male nor female* (Galatians 3:28).

The Christian body is both unregenerate and regenerate— the old man is unregenerate, the new man is regenerate; the old man is the natural man, the new man is the spiritual man; the old man is in the natural body, the new man is in the spiritual body; the old man is the outward man, the new man is the inward man; the old man is of the seed of Satan, the new man is of the seed of Christ.

The old man is absolutely distinct from the new man. There is no old man residue in the new man, and there is no new man residue in the old man. The one is unjust, the other just; the one is sinful, the other sinless; the one is strong, the other weak. The old man has the advantage over the new man, as the old man is an adult compared to the new man—a babe.

Between these two persons in one body, the battle rages. There is never a moment when the conflict ceases; no rest for the weary, no retreat for the downtrodden, no enemies taken, and no surrender for the vanquished. John Bradford captures the essence of the two natures of the Christian man, *just* in his sDNA and *sinful* in his nDNA, and of the ensuing, deadly battle that continuously waxes and wanes:

> One man therefore which is regenerate well may be called always just, and always sinful: just in respect of God's seed and his regeneration; sinful in respect of Satan's seed and his first birth. Betwixt these two men therefore there is continual conflict and war most deadly; "the flesh and the old man" fighting against "the Spirit and new

man," and "the Spirit and new man" fighting against "the flesh and old man." Which "old man" by reason of his birth that is perfect doth often for a time prevail against "the new man," (being but as a child in comparison), and that in such sort as not only others, but even the children of God themselves, think that they be nothing else but "old," and that the Spirit and seed of God is lost and gone away: where yet notwithstanding the truth is otherwise.[2]

Allow me, if you will, one last time to try to convince you, one last attempt to indelibly imprint upon the mind of your consciousness, of the only plausible explanation of why every Christian living today, is a chiral person; one who is in possession of two natures; two diverse sets of DNA; and two physical bodies—of one who enjoys to sin and of one who deplores sin, yea, who cannot sin! Listen again to the apostle Paul as he fights the battle that rages within his own body:

For that which I do I allow not: for what I would, that do I not; but what I hate, that do I. If then I do that which I would not, I consent unto the law that it is good. Now then it is no more I that do it, but sin that dwelleth in me. For I know that in me (that is, in my flesh,) dwelleth no good thing: for to will is present with me; but how to perform that which is good I find not. For the good that I would I do not: but the evil which I would not, that I do. Now if I do that I would not, it is no more I that do it, but sin that dwelleth in me. I find then a law, that, when I would do good, evil is present with me. For I delight in the law of God after the inward man: But I see another law in my members, warring against the law of my mind, and bringing me into captivity to the law of sin which is in my members. O wretched man that I am! who shall deliver me from the body of this death? (Rom. 7:15–24).

Dear Christian friend, lay claim to your salvation, enjoy the benefits afforded to you as recorded throughout Scripture. Dwell and contemplate on the nuances of "incorruptible":

"Incorruptible seed"—seed that will not corrupt itself, nor be corrupted by others. The seed deposited in the angel in his creation was corruptible, and the proof is that it corrupted itself. The seed lodged in man in his first make was corruptible, and the proof is that it was corrupted by others. But the seed planted in the believer in his regeneration is incorruptible as the seed of God—it neither corrupts itself nor suffers itself to be corrupted by others, it neither breeds evil nor catches evil. "Whosoever is born of God doth not commit sin; for His seed remaineth in him: and he cannot sin, because he is born of God." What is of the flesh in him sins; but what is of God in him is above sin.[3]

THE ARMOR

We are in a battle, a spiritual warfare between nDNA and sDNA, between good and evil that is played-out in the human genome. Ultimately, it is a spiritual battle between God and Satan. Battles are won by an army that is prepared and equipped for the battle. The army needs to know about the enemy and what are his tactics and capabilities. The army needs to know how, when, and where the enemy will attack and with what weapons he will use. The army must know what weapons they have available and how to use them, ere victory will elude a would-be victor. Beware! Satan will try to keep you from reading your Bible as pertaining to the tools necessary for victory, as it is your Spiritual Warfare Instruction Manual. It covers everything you need to know to obtain victory. Read it, study it, meditate on it day and night; and above all, obey it:

Put on the whole armor of God, that ye may be able to stand against the wiles of the devil (Eph. 6:11).

THE DEFENSE TEAM

The defense of any nation is paramount to its continued existence. As a nation, it must be in a constant state of preparedness, ready in a moment's notice to mount a defense against an attacking enemy. Attacks may come in a variety of forms and may originate from within or from without. A nation unprepared to defend itself is a nation living in constant fear of becoming a non-nation, a nation of poverty and servitude, of becoming a satellite of a more advanced military nation. There should be no excuse for being unprepared as the very life of the nation depends on it.

A high state of technological advances, an economy that is envious by everyone, and an advanced culture and society, are some of the factors that an enemy deems desirable to conquer in an advanced but unprepared nation. But are nations the only participants in a war for the very survival of their way of life? Is the only battlefield, where the conflict is fought, on the ground, on the sea, or in the air?

As do nations, individual people also have a keen sense of defense for their bodies from various forms of attack from within or from without. Who among us do not worry about our general health and well-being, of bodily injury that may well maim for life, or of diseases that seem to be everywhere just waiting for the right moment to attack? Our society has spent untold billions of dollars to construct numerous hospitals and health facilities to care for the ill, and staffed them with qualified doctors, nurses, and attendants in order to care for its people. This is no other than a defense system to combat particular forms of attack.

Drilling down deeper, our bodies also exhibit a keen sense of defense against attacks. Unfortunately, many of us are unaware of what is constantly going on just beneath the skin. It is there that the body is in constant conflict with all sorts of enemies; those with only one desire in mind; and that is to find a warm, comfortable place to live in, and to feed off of the sweet flesh of its host—you!

THE IMMUNE SYSTEM

Your defense system, in constant readiness, is referred to as the *Immune System,* a constellation of amazing molecules in possession of astounding intelligence and fortitude. They are in battle array, ready to defend, to attack, to kill, and to eat. The enemy in this case is not physical accidents or gunshot wounds, but is much more resilient and prepared. It is not fought on land, sea, or in the air, but in your body. This enemy consists mainly of diseases such as viruses, bacteria, and infections, caused by a host of microscopic organisms. The enemy is everywhere round about, in the food we eat, in the water we drink, in the bed we sleep in, in the mouth we talk with, and in the nose we breathe with. We just cannot get rid of them. They are always with us. They are in our blood.

THE LINEAGE

There are two main fluidic systems in the body; blood and lymph. The blood and lymph systems are intertwined and they have the responsibility of transporting the constituents of the immune system throughout the body. Blood flows from the pumping station of the heart into arteries, flowing into capillaries to bathe *every cell* of the body, and finally returns to the heart through veins to start the cycle over again.

Blood cells are manufactured by stem cells located in the bone marrow. Stem cells differentiate into several precursors that ultimately differentiate further into three specific types of blood cells: erythrocytes or red blood cells (RBCs); leukocytes or white blood cells (WBCs); and thrombocytes, referred to as platelets. The WBCs are further divided into two main lineages; the myeloid lineage, and the lymphoid lineage. There are about 10,000 WBCs per mm^3 in our blood supply and they live for about 9 days. RBCs total about 5 million cells per mm^3 and reproduce about 2.5 million new cells each second and live for about 120 days, after which they migrate to the spleen where they are phagocyted by macrophage; that is, eaten alive!

The immune system of the body is a prime example of this never-ending struggle for supremacy, and may very well be the primary location where the battle is fought. There are primarily nine different types of cells that function within the immune system (there are more), each having a different function to perform, and are assuredly associated with the health of one's body. Our immune system protects us from diseases and is a truly amazing constellation of responses to attacks from the inside or from the outside of the body. It has many facets, a number of which can change to optimize the response to unwanted intrusions. The system is remarkably effective.

The immune system is characterized by its ability to generate *antibodies* or *immunoglobulin* (the good guys), that can recognize and direct the immune system's mechanisms toward a vast array of different pathogens (the bad guys). When the immune system reacts toward antigens that are present on viruses, pathogens or foreign tissues, these processes are beneficial for the host because they lead to the elimination of the infecting agent.

What are the components of this highly sophisticated defense system? What are the names of the front line troops, where do they patrol in the body, and how do they get from the campground to the battleground? Do they talk to each other, do they remember what they did in the battle, and what exactly do they do when they fight?

First of all, the main effort of this amazing defensive army is to disallow enemy cells from entering the body. Cells on the outside of the body bother us not, but to allow one enemy cell to enter into the body and lock-on to a host receptor is equivalent to *firing the first shot heard around the world!* Immune cells fight against enemy cells that have breached the *walls.* They have overcome the natural barrier of the skin and entered the body, or more easily entered in through the mouth, nose, ears, eyes, or through any of the natural passageways.

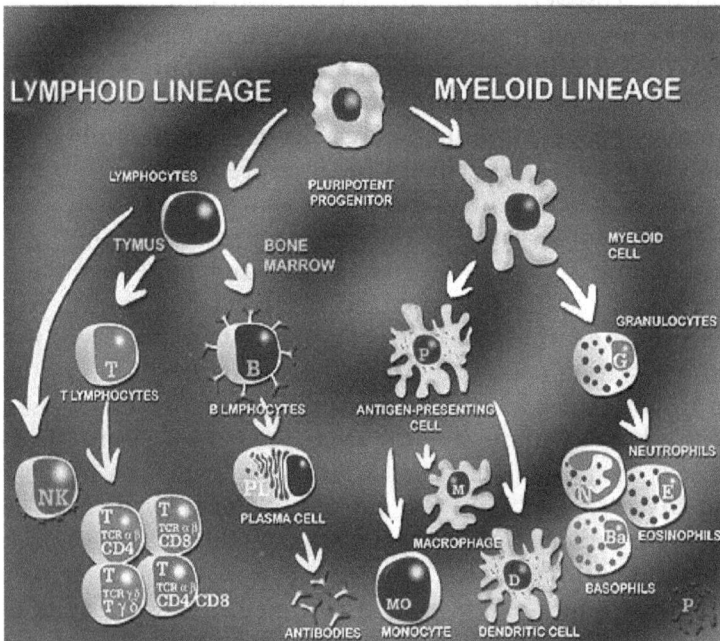

Figure 13.1: White Blood Cell Lineage

This military service is extremely disciplined, extraordinarily intelligent, and uniquely organized. First and foremost, at the top of the Military pyramid are the Pluripotent Stem Cells, Figure 13.1[4], that are made by the body in the marrow of the bones. These cells have the ability of differentiating into several types of cells with specific functions that are associated with different parts of the body. The pluripotent stem cells spin off a Myeloid Lineage and a Lymphoid Lineage, equivalent to establishing the Army and the Marine Corps. At the bottom is where the front line troops are located.

MYELOCYTES
White blood cells contain six of the nine elements of the immune system and each is assigned a specific task to do:

• Neutrophil
Neutrophil is a phagocyte, a "professional eating machine," the most abundant, the most active, and the most phagocytic of the white blood cells. The first phagocyte a pathogen is likely to encounter is a neutrophil. They respond to four keywords; "smell, seek, kill, and eat." These cells find the enemy by recognizing chemicals produced by bacteria in a cut or scratch, and migrate toward the smell. Upon reaching the site, they have the bacteria for a tasty lunch. They are really deadly.

• Eosinophil
Eosinophil is a phagocyte similar to neutrophil but less in abundance and activity, attacking parasites such as worms.

• Basophil
Basophil is a phagocyte similar to neutrophil and eosinophil but the least of the three. They circulate and can be recruited out of the blood into tissue when needed.

• Monocyte
A large phagocytic white blood cell circulating in the blood, which, when it enters tissue, develops into a macrophage.

• Macrophage
Macrophage (*big eater*) cells are large and versatile immune cells that act as microbe-devouring phagocytes and as antigen-presenting cells. They expose molecules of digested bodies and present them to more specialized cells.

• Dendritic Cells
Dendritic cells are highly efficient antigen presenting cells. They typically use threadlike tentacles to enmesh antigens,

break them down into fragments so that other cells can recognize them and then generate a specific immune response.

LYMPHOCYTES

Lymphocytes are the front line troops of the immune system, which is a defense against the attack of pathogenic microorganisms such as viruses, bacteria, and fungi. The three specific components of the lymphoid lineage cells are:

• B Cells

B cells perform the role of immune surveillance, and plasma and antibodies are its products. Each B cell comes with a unique immunoglobulin receptor on the surface of its skin that will bind to one specific antigen that displays the complement of the B cell's receptor. When a B cell encounters its triggering antigen, it gives rise to many large plasma cells. Each plasma cell is essentially a factory for producing one specific antibody, commonly known as an immunoglobulin.

• T Cells

T cells directly orchestrate immune defenses. They are divided into three categories: T_c (cytotoxic), T_h (helper), and T_s (suppressor). They retain antibodies on their skin membrane and use them to recognize infected cells. Cytotoxic T_c cells kill body cells infected by viruses or mutated by cancer cells. Helper T_h cells are needed to activate both B and T_c cells. Suppressor T_s cells turn off antibody production and call off the immune attack whenever the enemy is defeated.

• Natural Killer Cells

Natural Killer (NK) cells are another type of lethal lymphocyte. Like T_c cells, they contain granules filled with potent chemicals. They are called "natural" killers because they, unlike cytotoxic T_c cells, do not need to recognize a specific antigen before swinging into action. Both T_c cells and NK cells kill on contact. The killer binds to its target, aims its weapons, and then delivers a lethal burst of chemicals that produces holes in the target cell's membrane. Fluids seep in and the cell bursts.

CHAPTER THOUGHTS

What do all these pesky little things do? Well, they smell out, find out, kill, and eat attacking enemy cells. Cells eat cells.

There are about 100 trillion cells in the human body, where huge battles are constantly going on; enormous feasts, where everyone is invited; where everyone participates; and where everyone leaves full. You talk about a massacre!

Lymphocytes can remember every antigen encountered because they have a very good memory. When lymphocytes encounter an antigen for the second time, they respond quickly, vigorously, and specifically to that particular antigen because they remember what happened at their first meeting. This specific immune response is the reason that we do not contract chickenpox or measles more than once in our lives.

But there is one battle within the human body that we are mainly interested in, a very specific battle, a battle that has never entered into the mind of many believing Christians. Our Creator has prepared our bodies for this battle. Scripture admonishes us that this battle is for the very soul of a person. It is a battle that determines the way we live, the choices we make, and ultimately of where we will live for all eternity.

There can be nothing in our nature that does not find its source in the DNA molecule. When Adam sinned, he lost his spiritual image of God, lost communication, lost fellowship, and lost his ability to pray to his Father. Man now is depraved in this condition, not able to save himself or willing to be saved. He can do nothing but sin, as sin is now found in every cell of the human body. It has taken over full control of every cellular function, and, thoroughly entrenched, will not abrogate its position and rule.

Man is the sum total of all his cells. From the moment of conception to the moment of death, there is not one iota in the human body that is not cellular. Sin permeates everywhere, constantly raising its ugly head in every occasion; expressing and communicating its lethal desires to every other cell, causing the body to commit more heinous sins against the Lord.

Sin is the real enemy, an infectious disease, a formidable pathogen. It is ruthless and relentless, hostile and unremorseful, unforgiving and deadly. Sin affects the entire body, a formidable foe to the entrance of the holiness of God. The divine image that was lost due to the intrusion of sin must be regained. This is not a work for human ability to perform but must be accomplished by divine intervention.

Thank God for immune cells, for without them, we would be unable to survive. From the first neutrophil down to the last natural killer cell, we are protected from a horde of infectious diseases that would gladly have us for a tasty lunch.

But in the Christian body as we have defined it to be, there are natural immune cells, nImmunes, and there are spiritual immune cells, sImmunes, acting upon the body; in a similar fashion but are in no way apologetic toward each other. The intelligence latent in the immune system components are enormous, downright mind-boggling. Who can fathom the constant readiness, the resiliency, the prowess, the faithfulness, the organization, and the infinitesimal size and weight of the least of these warriors? Their main function is to protect the totality of the human body, a *gazillion* times its size, from infectious diseases and to do so with no advanced knowledge, no previous encounters, with no heart, no liver, and no brains! From their inception they never sleep but are on constant patrol. They have no time to think about a plan of attack, no understanding of how to identify upwards of one hundred thousand individual types of pathogens, and then to select one antibody out of a total of two million in its repertoire that will do the job of protection. What an amazing system!

14

LOVE, JOY & PEACE

"But the fruit of the Spirit is love, joy, peace...."
(Galatians 5:22)

E motions seem to lie at the root of the tree. Emotions are the state of a person, a mental and physiological state associated with a wide variety of feelings, thoughts, and behaviors. Humans cannot exist without emotions—examples of which are fear, anger, joy, love, hate, and jealousy.

Emotions are generally regarded as an indicator of physical and mental health, and appear to play a central role in many human activities and in the total well-being of all individuals. Wherever I see decay, brokenness, and dysfunction, I attribute it to the corrosive power of Adamic sin. From macrocosm to microcosm, from ecosystems to DNA, the effects of sin pervade us all. Given this perspective, there is no aspect of life that is not rooted ultimately in the supernatural.

I would suggest that brain chemistry and genetic structures are *vehicles* through which the Holy Spirit, via the "fruit of the Spirit," affects both our internal and external behavior. Conversely, Satan could, via genetic or neural alteration, stimulate a person to manifest the "works of the flesh."

> Based on our present knowledge of genetics and neurochemistry, it seems legitimate to posit that the Holy Spirit can affect spiritual growth--or produce the "fruit of the Spirit" at least in part by changing genes or the levels of certain chemicals in the human brain. . . . On some level, there has to be a meeting or interface of the physical and the supernatural. If we believe that the Holy Spirit is within the believer producing behavioral fruits such as

peace and joy, how does this occur? Such transformation appears to be mediated through a physical substrate to some degree, and research seems to show that genes and brain chemicals are such substrates.[1]

MOLECULES OF EMOTIONS

Candace Pert coined the phrase *Molecules of Emotion* in her award-winning book: *"Molecules of Emotion: Why You Feel the Way You Feel".*[2] She showed scientifically that certain molecules are related to emotions and that *neuropeptides and their receptors* are at the core of this fascinating science.

Neuropeptides enable systems to send signals to one another. They slip into receptors on cell surfaces and turn-on relevant processes. As emotions fluctuate, peptides sweep throughout the body in response, signaling organs and systems to perform myriads of physical changes.

Neuropeptides are transmitted throughout these systems via body fluids of blood, lymph, cerebral spinal fluid (CSF), and nerve pathways, as well as from neuron–to–neuron in the brain. No longer are they confined to the traditional neuronal circuits of the central nervous system.

The language of the scientific community is beginning to identify the human entity as a body–soul–spirit–Holy Spirit interface—where all emotions now emanate from the Holy Spirit to produce spiritual physiologic responses throughout the body. No longer are we to pay tribute to our natural feelings that run rampant in our natural nDNA flesh—to wrong emotions and hurtful physical responses—but are admonished to obey the dictates and leading of the Spirit into new and lively emotions and feelings.

God is a God of emotions! Is it not of a truth that this very same system, which gives billions of brain cells the media for communication one with another, provides also the means of communication with the spiritual realm emanating from without the physical body; an *involuntary, external system of applied emotions?*

THE FRUIT vs. THE WORKS

The *fruit of the Spirit* (Galatians 5:22–23) is a major factor in such a system and may be classified into three main groupings. The first group, the first three of the fruit—*Love, Joy,* and *Peace*, thoughts produced by God the Father; the second group—*Longsuffering, Gentleness,* and *Goodness,* thoughts produced by the Son; and the last group—*Faith, Meekness,* and *Temperance,* thoughts produced by the Holy Spirit—flow uninhibited throughout the Christian body.

Over against the three groups of the fruit of the Spirit are seventeen of *the works of the flesh* (Galatians 5:19–21), alias *the works of sin,* virtual antigen/pathogen complexes of evil that permeate the body. These seventeen have also been similarly combined into three distinct groups in order to show the "night and day" differences in the lives of believers and non-believers. Non-belief ushers in, and delights in, natural acts of: *Hatred, Envyings,* and *Strife,* inflicted by Satan (the Dragon, Rev. 12:7ff); *Uncleanness, Seditions,* and *Witchcraft,* inflicted by the Antichrist (the First Beast, 13:1ff); and *Idolatry, Wrath,* and *Revellings,* inflicted by the False Prophet (the Second Beast, 13:11ff). The eight remaining works have been distributed among the three groups as they are similar in nature.

The fruit of the Spirit is, in part, the *image of God* in the spiritual man, and the works of the flesh are, in totality, the *image of Satan* in the natural man. Either we respond to the joys of life with "the fruit" or to the fatalities of death with "the works." The battle array is certain—the fruit of the Spirit vs. the works of the flesh:

> *For the flesh lusteth against the Spirit, and the Spirit against the flesh: and these are contrary the one to the other* (Gal. 5:17).

From this response, it is interesting and increasingly evident, that each of our *spiritual* neuropeptides, emotions, and immune components, bear a stark relationship to the *divine* fruit of the Spirit; where we see our every thought and response of the body acutely tuned to the fruit. We describe this fruitful work of the Spirit as analogous to the spiritual formula:

$$\textit{The fruit of the Spirit} \rightarrow \textit{sNeuropeptides} \rightarrow \textit{sEmotions} \rightarrow \textit{sImmune Cells},$$

where the moral excellencies of God are incorporated into the body by the Spirit through thought processes, producing a plethora of spiritual sNeuropeptides that feed upon the fruit as a substrate, i.e., the fruit of the Spirit → sNeuropeptides:

> The nine *fruit of the Spirit*—Love, Joy, Peace, Longsuffering, Gentleness, Goodness, Faith, Meekness, and Temperance can be viewed as the theological substrates of the nine, biological new birth *spiritual Amino Acid Neuropeptides*—sOxytocin, sSerotonin, and sEndorphin; sGABA, sDopamine, and sEpinephrine; and sVasopressin, sAnandamide, and sOrexin.

Webster defines the meaning of *substrate* as: *"Any stratum lying underneath another,"* an underlying, supporting layer, as soil lying underneath a flowering plant—the plant feeding upon the nutrients in the stratum. Extending this into the spiritual realm: *Eternal Life* and *Eternal Death* are the substrates for the *elect* and the *non-elect*, supporting layers for the growth of all that is *good* or *evil* in the hearts of men; or, as the *shed blood of Christ* is the substrate of the *new birth conception*, an underlying, supporting layer for the growth of the Christian zygote.

The spiritual formula also elicits spiritual sEmotions that in turn feed upon sNeuropeptides as substrates that flow throughout the body, i.e., sNeuropeptides → sEmotions:

> The nine *spiritual Neuropeptides* are the biological substrates of the nine *spiritual Emotions* of sLove, sJoy, sPeace, sLongsuffering, sGentleness, sGoodness, sFaith, sMeekness, and sTemperance.

Extending the formula further, spiritual sImmune Cells in the human body feed upon sEmotions acting as substrates, i.e., sEmotions → sImmune Cells:

> The nine *spiritual Emotions* are the biological substrates of the nine *spiritual Immune Cells*—sNeutrophil, sEosinophil, and sBasophil Cells; sT Cells, sB Cells, and sNK Cells; sMonocyte, sMacrophage, and sDendritic Cells.

I believe, as many others do in the scientific field of biology, that there is a specific primary neuropeptide responsible for producing a specific primary emotion and feeling in the body.

There is, however, no current scientific data available to definitively identify such neuropeptides.

SPIRITUAL FRUIT: GROUP 1

We have, however, searched and arrived at what we believe to be—*nine spiritual neuropeptides* that produce *nine spiritual emotions* that relate primarily to *nine spiritual immune cells* of the body, all produced and powered by the divine *nine fruit of the Spirit!* The Love of God is the first of the fruit to be considered, producing spiritual sLove in our body:

- **LOVE → sOxytocin → sLove → sNeutrophil**
The love of God is shed abroad in our hearts by the Holy Ghost which is given unto us (Rom. 5:5).

THE LOVE OF GOD

The Love of God (Gk, a*gape*) is first in the long list of divine attributes of God. Agape is love that is of, and from, God. God does not merely love; He is Love. God loves because that is His nature to do so and the express image of His being. The Love of God is a self-sacrificing love, a love that first entered the world at Calvary where Christ gave His life for you and for me, a sacrificial love void of all envy and malice. We are the undeserving recipients upon whom He lavishes His Love.

But this type of love does not come naturally to humans. Because of our fallen nature, we are incapable of producing such a love. If we are to love as God loves, that love can only come from its true Source. It is clear that only God can generate within us this kind of self-sacrificing love, which is proof positive that we are His children. Because of God's love toward us, we are now able to love one another with agape love.

The first fruit in the list is love and the last is temperance; a pair of bookends that bind the remaining fruit in their respective order; a love that surpasses human understanding and causes a person to strive for the fullness of God, and a temperance that exercises control over the whole of the Christian body. The following anonymous lyrics express these thoughts precisely:

What wondrous love is this, O my soul, O my soul!
What wondrous love is this, O my soul!
What wondrous love is this, that caused the Lord of bliss

To bear the dreadful curse for my soul, for my soul
To bear the dreadful curse for my soul!

• sOxytocin

The agape Love of God finds its way into our emotions and feelings. We feel good when we love someone. His thoughts of agape love, impressed upon the frontal cortex of the brain by the Holy Spirit, produces and releases neurotransmitters that migrate within the brain to the *hypothalamus,* and in turn to the *pituitary gland,* and then on to the *amygdala.*

This HPA complex, under agape Love attack, produces and releases the neuropeptide *oxytocin* that directly affects several organs and systems, one of which is forming everlasting bonds between mother and child. The amygdala, however, opens up the floodgates for the nine amino acid oxytocin neuropeptide to flow uninhibited up and down the love pathways of our body.

The amygdala, the seat of human emotions, whose primary role is in the processing of emotional reactions, upon the proper stimulus, receives and releases oxytocin; not through "hardwired" pathways to specific organs as does the pituitary hormone, but into the fluidic systems of the body; through the blood and the cerebral spinal fluidic systems; impressing upon specific receptors feelings of affection and brotherly love; and is the main ingredient in the formation of everlasting bonds between God and man; spiritual bonds; bonds that will thrive beyond eternity; hence the designation—*the love peptide.*

• sLove

Spiritual love in the life of a believer is an emotion, a moral excellence, portraying but a small portion to the world of the divine character of God; a small measure of His infinite love. We will adhere to the following definition to state unequivocally that the items listed in the fruit of the Spirit, love, joy, peace, etc., is the driving force behind spiritual emotions. Everything has its roots in emotions. We are creatures of emotions and feelings; therefore, the fruit enumerated by the Spirit is basically and fundamentally emotional.

Unlike our English word for love, agape is not used in the Bible to refer to romantic or sexual love, nor does it refer to close friendships for which the Greek word *philia* is used. Agape love is unique and is distinguishable by the nature and character of the one who is in possession of it.

The emotion of spiritual love, sLove, emanates from the Spirit and holds the higher emotional charge. Love cures people, both the one who gives it and the one who receives it. Loving and hating are not just emotional states but both have direct physical effects on the body. Just as feelings of hate raise blood pressure, shorten the breath, and create a psychological distance between people; feelings of love, in contrast, decrease blood pressure, lengthen and slow breathing, and blur the distinction between one person and another. Not only does hate hurt the one dispensing it, but it hurts the person feeling it; while love benefits both the giver and the receiver.

While oxytocin acts as an *agonist* for the emotion of spiritual love, it also functions as an *antagonist* for the infusion of the natural emotion—*nHatred*—the antithesis of sLove; the first work in an unbeliever's arsenal in "the works of the flesh," directed toward his god, Satan, that vies for the very same receptors. In this wise, hatred, a sibling of the original antigen/pathogen complex encountered in the Garden of Eden, is done away with as specific love receptors are occupied by the peptide sOxytocin; prohibiting hatred from gaining entrance into the same cells to spread its venom. The old adage runs true: The more you love the less you hate; for *Hatred stirreth up strifes: but love covereth all sins* (Proverbs 10:12).

Agape love is the highest obtainable in the arsenal of the Spirit. As the elect worship and come to the communion table to partake of the bread of His body and sip of the cup of His blood; and as they confess before a righteous and holy God to receive forgiveness and be cleansed from all sin; feelings of love for Him wherein He first loved us, of joy of being set free from all evil, and of peace with God, swell within the body, for: *The blood of Jesus Christ his Son cleanseth us from all sin* (1 John 1:7), setting in motion an avalanche of molecules of emotion, feelings of love that abound in the spirit and the soul.

• sNeutrophil

The nine fruit of the Spirit is not a natural human trait at all but is the express character of Christ Himself. Those to whom it is given demand a higher immunological response from the nine white blood cells of the immune system; a higher immunity than that possessed by the non-elect; a spiritual sImmunity for the elect Christian and a natural nImmunity for the non-elect unbeliever.

Sin, however, is the ultimate stressor to the body that presents itself as an external antigen/pathogen complex of natural hatred, *nHatred,* and attempts to issue its venom to the internal parts of the body. The "works of the flesh" that honors Satan, spread throughout at the speed of light. The natural nImmunological system is overwhelmed, powerless to resist, and ultimately unable to respond.

In contrast, the sImmune system is constantly at a high level of preparedness against such attacks and immediately mobilizes for the main thrust. sNeutrophil, a phagocyte, a professional eating machine, is the first to respond and the first to arrive at the site of attack. From this point on, there is no mercy shown and no enemy taken. nHatred is no match for the defensive action of sNeutrophil, hence, the new birth is forever preserved, untainted, sin free, enjoying infinite health.

• JOY → sSerotonin → sJoy → sEosinophil

These things have I spoken unto you, that my joy might remain in you, and that your joy might be full (John 15:11).

THE JOY OF GOD

The Joy of God (Gk. *chara*) stands as a monument next to the Love of God and comes to us because of His presence within us. The divine attribute of Joy in the essence of God is unspeakable. You cannot begin to describe it, nor can you obtain it for yourself. His Joy is not our joy—our joy is derived from His Joy. He freely bestows His Joy by means of His Grace to those who believe.

Joy is only one aspect of His character, not only of what He gives but it is also that which He experiences. God seeks to share His joy with us because He is joyful, expressing joy in what His hands have made, of all that He has created. He is the source of all love and the blessing of all joy.

• sSerotonin

The Joy of God produces the spiritual neuropeptide serotonin—*the joy peptide,* that is vital to the chemistry of the natural body and to the feelings of joy in the human heart. One either suffers debilitating bouts of deep depression with low levels of the peptide, or on the other hand, will abound with joy unspeakable with appropriate levels of serotonin.

Depression and anxiety, aggression and hopelessness, worry and confusion, are inhibited as optimum levels of the peptide are maintained throughout the body. The messages of God contained in sSerotonin literally fly across the synapses to energize the next neurons on their way forward in the joy and bliss pathways of the human brain, and then find their way down through the body; happening as fast as thought and life itself. Whenever we think, or feel, or respond to a physical change or sensation, neurons are firing sSerotonin peptides; generating new thoughts, new feelings, and new sensations:

> Serotonin is the most important chemical in our bodies. You can't eat and digest serotonin, and you can't get it through an injection. God created us so that only the brain can make serotonin, and it floats in the synapses (spaces) between our 40 billion nerve and brain cells. . . . If you have the right amount of serotonin in your brain cell synapses, you are filled with love, joy, peace, longsuffering, gentleness, meekness, energy during the day, great sleep at night, and less physical pain.[3]

• sJoy

Whom having not seen, ye love; in whom, though now ye see him not, yet believing, ye rejoice with joy unspeakable and full of glory (1 Pet. 1:8).

Spiritual joy is the infallible sign of the presence of God. Greater than all our penances and sacrifices is to live joyfully because of the knowledge of His love, His grace, and His forgiveness. Knowing this, there is a certain joy that comes to the soul, a relief from the needless worry associated with the "works of the flesh."

Be joyful! Experience the calm delight that flows throughout the body, lighting up the dark recesses of the heart. Cheerfulness and gladness are the marks of a Christian at the level of spiritual joy, *for the joy of the LORD is your strength* (Nehemiah 8:10). sJoy is a joy unto salvation and the knowledge of it sets us free from the past, gives faith for the present, and a glorious hope for the future.

• sEosinophil

As in neutrophil, eosinophil is a phagocyte that delights in devouring disease-carrying pathogens in the natural body.

Eosinophils primarily deal with parasitic infections and are also the predominant inflammatory cells in allergic reactions. The most important causes of eosinophilia include allergies such as asthma, hay fever, and asthma.

In the spiritual body, sEosinophil inhibits *nEnvyings*—the antithesis of sJoy, the second work in "the works of the flesh," aimed at pleasing Satan, adding to the arsenal of weapons— active soldiers on guard, ready to spring into action at a moment's notice to maintain perfect health.

• PEACE → sEndorphin → sPeace → sBasophil
Peace I leave with you, my peace I give unto you: not as the world giveth, give I unto you. Let not your heart be troubled, neither let it be afraid (John 14:27).

THE PEACE OF GOD
The Peace (Gk. *eirene*) of God, beautifully rendered in the refrain lyrics by Warren D. Cornell of that old time favorite hymn, *Wonderful Peace*, speaks volumes of the Peace that flows from the throne of God that fills the heart of man:

> *Peace, peace, wonderful peace,*
> *Coming down from the Father above!*
> *Sweep over my spirit forever, I pray*
> *In fathomless billows of love!*

• sEndorphin
The Peace of God produces positive levels of endorphin peptides, reducing pain while increasing the level of euphoric bliss of happiness and pleasure; thus one feels at peace within himself and with the world; hence the designation—*the feel-good peptide*, the body's natural solution to stress.

Endorphins act by locking onto receptors in the nervous system that transmit pain messages to the brain. Once the endorphin, the "key", docks in the "lock" of the receptor, pain-causing chemicals are prevented from transmitting their messages. Thus, to stop pain signals from reaching the brain and causing discomfort, you could eat certain foods, such as chocolate and chili peppers that lead to enhanced production of endorphins. Laughter is another option. Strenuous exercise and massage therapy can also activate endorphin production to reduce pain; but a much simpler and easier way to obtain peace is through the Spirit.

• sPeace

Spiritual peace is not the absence of conflict, but the presence of God no matter what the conflict; it is a gift of God. He bestows peace by the Word of His power, a lesson learned by observing the growth of the fruit tree. The kind of fruit that grows on the outside of the tree is a reflection of the nature of the tree inside. Apples grow on apple trees and pears on pear trees, but the fruit of the Spirit that grows in your life is an outgrowth of the nature within. It is the new man, the new life, the new nature, that expresses itself in the kind of fruit that grows on the outside for others to see and to emulate.

sPeace (Heb. *shalom*), the third of the fruit of the Spirit, rounds-out the trilogy with love and joy, which portray our thoughts toward God. It carries with it the notion of wholeness and tranquility even in the face of adversity and chaos. The "works of the flesh"—*nStrife*—the antithesis of sPeace, the third of the works in this group that honors Satan, is no match for sPeace, for Jesus is *The Prince of Peace* (Isaiah 9:6); Satan—*Beelzebub the prince of the devils* (Matthew 12:24).

• sBasophil

Basophil contain granules filled with chemicals that are toxic to other cells, are used to destroy invading bacteria and other pathogens, and are involved in inflammatory responses and allergic reactions. They destroy pathogens by engulfing them and then destroying them by releasing their toxic granules. Basophil releases histamine that promotes an increase of blood flow to tissues, and also heparin which prevents blood from clotting too quickly.

CHAPTER THOUGHTS

Scientists have spent enormous resources attempting to identify what immune cells have what immunopeptide receptors located on their plasma membrane. For instance, does the neutrophil immune cell contain oxytocin receptors on its surface that will surely attract the "love peptide," setting in motion an avalanche of beneficial immune interactions throughout the body? In like manner, do eosinophil and basophil cells contain serotonin- and endorphin-type receptors respectively? The answers to these life-transforming questions should be on the mind of all scientists simply because of the tremendous health advantages afforded to the general public.

Imagine what medicines could be developed for the cure of cancer, AIDS, the flu, and even the common cold. Any wonder then what the Christian body will be like when we all obtain infinite health in the world to come. Even then, you couldn't even come close to the advantages associated with spiritual neuropeptides, emotions, and immunes flowing uninhibited, that will surely display our Christ-like character.

Almost all neuropeptides found within the Central Nervous System are also located in and on immune cells. They make the same chemicals as the brain makes when it thinks. An immune cell has emotions; it has desires; it has intellect; it knows how to remember. The immune cell is a thinking cell.

When you think a thought, you make a molecule. Thoughts and emotions translate into peptides in the brain. This dialect is not in English with a Swahili accent, but in the precise language of neuropeptides. Nothing that you say to yourself, even in your sleep, escapes the attention of immune cells. It's a conscious being—you have *gazillions* of these little creatures talking to themselves all day long.

Every cell in our body thinks. Every cell is actually a mind. Every cell communicates with every other cell. We have a body-mind located everywhere, a bidirectional communications system between the immune system and neuropeptides in the brain. The body is an amazing design. At the cellular and molecular levels, and then down through the atomic level and beyond—into levels of life unknown—God has done wondrous things. I never cease to wonder how these little molecules function with a keen sense of awareness of their responsibility. How is it possible for an unconscious, lifeless fluid, without a brain, without eyes or ears, without knowledge, composed of a certain combination of atoms too small for the eye to see, to act in such an intelligent, organized, and timely way; how all of these chemicals, being "dead as a doornail," could ever determine who we are, how we think, how we act, and how emotional and healthy we are to be.

Know this—the *fruit of the Spirit* lodged within a believer permeates into the deepest recesses of the human soul.

15

THE NEW BIRTH BLASTOCYST: *THE LITTLE CHILDREN*

"I write unto you, little children, because ye have known the Father" (1 John 2:13)

The seed has been sown in Matthew 13. The *good seed* (v. 24) has fallen on *good ground* (v. 23). The *good seed* is the *other seed* (v. 8), seed that is different from the seed sown *by the wayside* (v. 19), *in stony places* (v. 20), among *the thorns* (v. 22), or the *tares* (v. 25). There is good seed and there is bad seed. There is good ground and there is bad ground. It is the good seed that yields good fruit from the good ground, and it is the bad seed that yields bad fruit from the bad ground:

> *Even so every good tree bringeth forth good fruit; but a corrupt tree bringeth forth evil fruit. A good tree cannot bring forth evil fruit, neither can a corrupt tree bring forth good fruit* (Matt. 7:17–18).

The *good seed* has sprung up. The *good fruit*, the spiritual zygote, a one-cell organism, now begins to grow into a multi-celled organism, receiving natural nutrients from its mother and spiritual nutrients from the Holy Spirit:

> As in the natural so in the spiritual: there is a "begetting", a "conception" or Christ being formed in the soul, a "birth" and that birth evidenced by a "cry", and the newborn babe desiring "the sincere milk of the Word"; so there are many features in common between the natural and the spiritual infant.[1]

SPIRITUAL CHILDBIRTH: PHASE 2

Following the act of effectual calling, the Holy Spirit instituted the act of regeneration to initiate the birth process. In order to bear spiritual offspring, you must first be called-out, regenerated, and then negotiate through a zygote. Reconciliation follows regeneration, producing a spiritual blastocyst in the body, i.e., *Reconciliation → Cellular → Blastocyst → Little Children* continuing the process of spiritual growth (refer to Figure 11.1).

RECONCILIATION

Reconciliation is the normal continuance of the act of regeneration, producing effects upon the Christian body in unison and in synchronism; the pronouncement that what regeneration has accomplished in the zygote, acceptance of the blastocyst into the family of God is effective and complete. The new Christian is at peace with God and no longer an enemy. The Psalms speak out concerning this relationship:

> *Mercy and truth are met together; righteousness and peace have kissed each other. Truth shall spring out of the earth; and righteousness shall look down from heaven* (Psa. 85:10–11).

Mercy and the truth of God met together when the righteousness of God embraced the peace of man. Where did these meet? In Christ Jesus. When were they reconciled? When He poured out his life on Calvary.[2] Permanent peace can exist only on a foundation of righteousness; earth and heaven meet in man's redemption.

CELLULAR

Continuing the analogy of a tree planted in good ground yielding good fruit, the development of the small sapling into a towering tree is a process veiled in secrecy. So also with our spiritual growth, it is unrecognizable, as we cannot perceive with our eyes the actual process. Trees grow gradually; some grow very slowly, while others reach maturity more quickly. The nature of trees is to move toward heaven, slowly lifting its head higher and higher, its growth measured by its *upward* progress. Attracted by the rays of the sun, the growth is matched by its becoming more deeply rooted in the ground, as is the growth of the *little children* deeply rooted in what the *babes* have firmly established as the seed, in its growth *downward* into the depths of the soul:

> Such upward growth will consist of stronger yearnings after God, more constant and frequent seekings after Him, a closer acquaintance with Him, a warmer love for Him, more intimate communion with Him, a fuller conformity to Him, and a deeper joy in Him.[3]

The more we are weaned from the world, the higher we grow, the deeper our desire for spiritual things, and the sweeter they become to our taste. As our love for Him increases, we place a higher esteem on Him, a greater delight in Him, and the more the heart pants after Him.

The single-cell Zygote continues to grow by expanding the number of cells, first to two, then four, then eight, then to sixteen, and then up to 100 where it is now referred to as a Blastocyst, another step in its process towards the new birth. The number of cells now has expanded according to the "budding-out" process of dNeuropeptides (d for Devine) from the original regenerated cells called-out for election.

A spiritual cell contains all the information and faculties that it will ever obtain. The cell does not continue to grow in wisdom and knowledge as time goes on, but are mature in all its faculties as at the day they were regenerated. They will not be any different from the spiritual cells of the resurrected body that will eventually inhabit heaven. The only problem is—at this stage—we only have a handful of spiritual cells! We are overwhelmed by natural body cells that have no affinity for spiritual realities. A mature Christian, on the other hand, is overwhelmed with spiritual cells that exert a great influence over the behavior of the remaining natural cells.

BLASTOCYST

The blastocyst is preceded by the zygote, the fertilized egg cell, and succeeded by the embryo. At this point, it continues to differentiate into two discrete cell masses, the outer cell mass (OCM) and the inner cell mass (ICM). The OCM, the trophoblast, will form the placenta and other supporting tissues needed for fetal development within the uterus, and the ICM will go on to form the embryoblast in Phase 3, giving rise to the tissues of the body. Figure 15.1[4] below is a virtual representation of a spiritual blastocyst. As in natural pregnancy, the ICM contains pluripotent stem cells that differentiate into specific organs or supporting systems for the spiritual body.

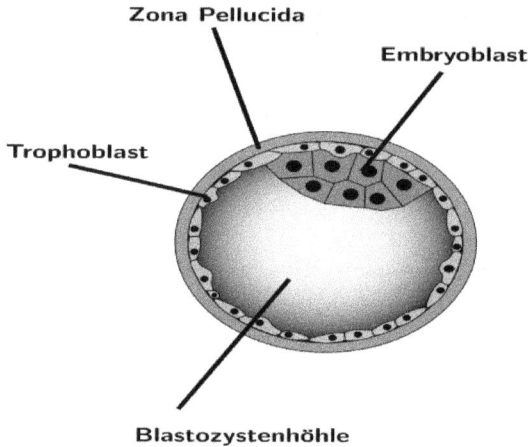

Figure 15.1: Blastocyst

Let us devise a scenario that may or may not be entirely accurate, bearing in mind that I, also, am human and fraught with errors. Upon conception, amazing procedures are set in motion. The natural womb, defined as the uterus of the human female, now is *not a singular womb* of a would-be mother waiting to give birth to her fetus, but is now defined as not belonging to one specific gender but to everyone born of the Spirit:

> The womb (clearly a *virtual* womb) is now exemplified by the entire spiritual body of the Christian (of both male and female), consisting, not as it is defined in natural pregnancy as one singular uterus, but of many uterus subsets, i.e., *uteri,* one individual uterus for each individual organ-specific system; infused throughout with spiritual, pluripotent, blastocyst, stem cells; each supported by a spiritual placenta, umbilical cord, and amniotic sack. Anything more than this, I do not understand (You will have to ask an OB/GYN!).

Admittedly, this is an approximate process but since the natural body at the general resurrection will be raised fully a spiritual body, then every natural nCell of the natural body must undergo a regeneration into spiritual sCells to produce a totally spiritual body, not one cell being omitted or wasted.

The trophoblast in the OCM in each blastocyst cell then seeks out each organ system that its pluripotent stem cells direct, and attaches to that system. Remember, the pluripotent stem cells contain all the information in the genetic code necessary to perform the necessary functions. There is not one organ system that is to be without an identical blastocyst cell impregnated. Each organ system then becomes an *organ-specific uterus* for the development of each individual spiritual organ pregnancy. Spiritual pregnancy consists of as many pregnancies as there are organ systems.

In this wise, spiritual pregnancy cannot be identical to natural pregnancy. Spiritual pregnancy is at war with the natural body, and therefore must invade *each* natural organ system of the host, the natural body, in order to transform natural nDNA organ specific flesh, to spiritual sDNA organ specific flesh, and to accomplish this on a cell-to-cell, organ-to-organ, and system-to-system basis.

All the while, the birth process continues. The organ specific, spiritual trophoblasts invade and penetrate the external lining of *each* natural, organ system uterus, simultaneously and in unison, and begin to grow within each host system, an invading sibling, each recognizing the other as invading pathogens. So now the war begins. Sound the alarm! Prepare for battle, for this process continues until every organ system in the natural body is *impregnated* with a spiritual blastocyst of the proper differentiated sCells suitable for that host organ.

Existing host arteries and veins in each organ system are now compromised and the placentas soon take form and become the boundary line between the natural host organ cells and spiritual blastocyst cells; pathways whereby natural nutrients are to be administered via the newly formed umbilical cords. (It was tough, but necessary, getting through all that!).

LITTLE CHILDREN

Young Christians are not to remain babes indefinitely but are admonished to *grow in grace, and in the knowledge of our Lord and Saviour Jesus Christ* (2 Peter 3:18). At this stage of its newly obtained spiritual life, blastocysts are classified as *little children* in Christ. The Apostle, writing in 1 John 2, addresses three basic groups of Christians, beyond those of *newborn babes* in the zygote phase, those who are at different stages in their Christian life as they continue to grow: *little children*—those at the blastocyst stage; *young men*—those who have progressed to embryo in Phase 3; and, *fathers*—those who have matured in the faith as a fetus in Phase 4:

> *I write unto you, fathers, because ye have known him that is from the beginning. I write unto you, young men, because ye have overcome the wicked one. I write unto you, little children, because ye have known the Father* (v.13).

A little child is a little child, but a little child is still *in* Christ! The term "little children" in verse 13 is from the Greek *paidion* and is different from the term "little children" as used by the Apostle to further identify the newborn babe, *brephos* (1 John 2:12), in Phase 1, where the Greek word used is *teknia*. The *teknia* little child is more immature than the *paidion* little child, consisting now at this stage of development of greater than 100 cells. As each new spiritual cell is added to each host organ-specific system, the uteri become larger and larger, each exhibiting new distinguishing marks of a Christian, in that they:

• Have known the Father
I write unto you little children, because ye have known the Father (1 John 2:13).

By knowing the Father, by having an intimate relationship with Him, by having embedded within himself the incorruptible seed *which liveth and abideth forever* (1 Peter 1:23), the Apostle can truly affirm that the *paidia* little child has been regenerated and is now reconciled to his Father by the blood of Jesus. The blastocyst little child has an understanding of God, though infantile, but yet so strong, that the knowledge of God is firmly written upon every spiritual cell in his body:

> It is a statement, which needs to be particularly taken to heart and pondered . . . for it plainly signifies that unless

we "know the Father" we are not entitled to regard ourselves as being His children . . . an acknowledgement—in
their infantile way—of their parents, aiming to call them
by their names ("papa" and "mamma") in distinguishing
them from others . . . to acknowledge God to be their *Father,* and this they do by expressing, in their way, their
attachment to Him, their delight in Him, and their dependence on Him, lisping out His name in their praises
and petitions before the throne of grace.[5]

Little children have known the Father, of who He is and of
what He has accomplished, but have little effect upon the
workings of the Christian body. They are immature to the extent that they are yet unknowledgeable in the Word; need to
grow further in grace, and thereby to anchor one's self to the
Rock of their salvation. Nevertheless, little children produce
various levels of spiritual fruit in their lives, relating to the
fruit that each stage of growth is able to produce:

*And these are they which are sown on good ground;
such as hear the word, and receive it, and bring forth
fruit, some thirtyfold, some sixty, and some an hundred*
(Mark 4:20).

First the blastocyst *little children, who have known the Father,* producing spiritual fruit *thirtyfold;* followed by the embryonic *young men, who overcome the wicked one,* with *sixty;*
and finally, the fetal *fathers, who have known him that is from
the beginning,* with *an hundred;* corresponding to the three
grades of growth—*the blade*—*the ear*—and *the full corn:*

*For the earth bringeth forth fruit of herself; first the
blade, then the ear, after that the full corn in the ear* (v.
28).

• Our Friend
*Henceforth I call you not servants; for the servant
knoweth not what his lord doeth: but I have called you
friends; for all things that I have heard of my Father I
have made known unto you* (John 15:15).

Once we were God's enemies, but now we are His "friends";
once we were in a state of condemnation because of our sins,

but now we are forgiven; once we were at war with God, but now we have peace through our Lord Jesus Christ. Dear Christian friend, can you imagine the Sovereign Creator and Ruler of the universe, the One who designed life and upholds all things by the word of His power, calling *you*—His *friend!*

Can you imagine that Christ died for us while we were still His enemies? Is there anyone else who would be willing to die for you, one who would give his life for you, who knew personally that you abhorred him? It is difficult to imagine the kind of love that would die for another, yet His love runs far deeper than we can imagine.

Christ reveals the Father's thoughts to His friends! Christ never forsakes His friends! Having given the clearest proof of friendship, Christ opens his mind to us; makes known his plans for us; and acquaints us with his death, his resurrection, and ascension. What greater friend could we ask for?

• A Chimera
For I know that in me (that is, in my flesh,) dwelleth no good thing: for to will is present with me; but how to perform that which is good I find not. For the good that I would I do not: but the evil which I would not, that I do. Now if I do that I would not, it is no more I that do it, but sin that dwelleth in me (Rom. 7:18–20).

The apostle Paul bemoans his inability to adequately respond to the tension constantly arising between his two combative natures. sBlastocyst cells exist in a non-harmonious relationship with its sibling, the natural body. We exist as a *chimeric* person, a person with two sets of different cells in every organ, natural cells with nDNA and spiritual cells with sDNA. This mixture of cells is uniform throughout the Christian body.

Historically, a chimera is a mixture of two or more species in one body and has its roots in Greek mythology, a fire-breathing monster resembling a lion in the forepart, a goat in the middle, and a dragon behind. It is a child of Typhon and Echidna. Echidna devastated Lycia in Asia Minor until she was slain by the Corinthian Bellerophon on the winged horse Pegasus. A tetragametic chimera matches a Christian chimera very closely:

This happens at a very early stage of development, such as that of the blastocyst. Such an organism is called a tetragametic chimera as it is formed from four gametes — two eggs and two sperm. Put another way, the chimera is formed from the merging of two nonidentical twins in a very early (zygote or blastocyst) phase. As such, they can be male, female, or hermaphroditic.

As the organism develops, the resulting chimera can come to possess organs that have different sets of chromosomes. For example, the chimera may have a liver composed of cells with one set of chromosomes and have a kidney composed of cells with a second set of chromosomes. This has occurred in humans, and at one time was thought to be extremely rare, though more recent evidence suggests that it is not as rare as previously believed. Most will go through life without realizing they are chimeras.[6]

Christians—Get used to it! You are a Chimeric person with two different sets of chromosomes, one of the natural and one of the spiritual, one you were born with and the other of Christ!

CHAPTER THOUGHTS

Not only are we, as Christians, regenerated to newness of life, but are now reconciled and are at peace with our awesome God. The blastocyst will continue to grow, and if not hampered by the Holy Spirit due to continued sin in the natural body, will overcome the entire body, where no trace of natural cells are to be found.

If you are a young blastocyst in your spiritual walk, know that God is your friend, having been in His loins before birth, and are a chimera-type of person: natural and spiritual.

Nothing now can hinder us. We are free at last! No need now to worry of life or of death; no need now for death has no hold upon us. Therefore, rejoice; I say, rejoice!

TIDBITS OF PART IV
- God is Love.
- God *bara* the Heaven and the Earth:
    o And it was Perfect.
- Satan—The Author of sin:
    o God permits it.

- Sin:
  - o Mutations in DNA sequences.
- WAR!
  - o sDNA vs. nDNA.
  - o The fruit of the Spirit vs. the works of Satan.
  - o Where?
    - ♦ In the body!
  - o Who?
    - ♦ The "new man" vs. the "old man."
  - o The defense team;
    - ♦ The immune system.
  - o The Linage;
    - ♦ White Blood Cells, the good guys.
  - o The Lymphocytes;
    - ♦ B–, T–, and NK Cells.
  - o The Myelocytes:
    - ♦ Neutro–, Baso–, Eosinophils,
      Monocyte, Macrophage, Dendritic Cells.
- Emotions of God:
  - o The fruit of the Spirit.
- FRUIT→sNeuropeptides→sEmotions→sImmune Cells:
  - o LOVE → sOxytocin → sLove → sNeutrophil.
  - o JOY → sSerotonin → sJoy → sEosinophil.
  - o PEACE → sEndorphin → sPeace → sBasophil.
- Love, Joy, Peace vs. hatred, uncleanness, idolatry.
- Phase 2: *Reconciliation → Cellular → Blastocyst → Little Children:*
  - o Reconciliation;
    - ♦ Righteousness and peace have kissed each other.
  - o Cellular;
    - ♦ As a tree planted in *good ground.*
  - o Blastocyst: Inner CM, Middle CM, Outer CM;
    - ♦ Inner: Endoderm → muscle, organ linings.
    - ♦ Middle: Mesoderm → Embryoblast → embryo.
    - ♦ Outer: Ectoderm → Trophoblast → placenta,
      epidermis & nervous system.
  - o Little Children;
    - ♦ Know the Father.
    - ♦ A Chimera.
- Each organ–system → an *organ–specific uterus.*
- Spiritual → Cell by cell, organ by organ, flesh by flesh.

PART V

The Battle for the Soul

16

LIGHT'S SHINNING GLORY

"God is light" (1 John 1:5)

We have identified that: 1) the *External Source of Empowering Energy*, the first requirement for reversal or suspension of the Second Law of Thermodynamics, is the First Person of the Trinity, God the Father, the Omnipotent, who creates out of nothing, *ex nihilo*, who possesses the power necessary to perform the required initial molecular transformations in the human body—through the action of metamorphosis—where natural nDNA sequences are converted to spiritual sDNA sequences, and where the plan of God for a new creation is infused into the mortal body of man; and 2) the *External Source of Conceptual Information,* the second requirement for suspension of the Second Law, lies resident in the Second Person of the Trinity, God the Son, the Omniscient, Jesus Christ, the Word of God, the True Light, in Him resides all information and intelligence, the fountain of all knowledge, the *Word of Life.*

Our metamorphosis is not a process of some shallow behavior modification program, but it is an actual biological change that takes place at the molecular level in the depths of our being; an instantaneous change in the nature of man; a change from one type of flesh into another type of flesh; of an old nature to a new nature. If the nature of a duck is to quack, the nature of the sun is to shine, and the nature of the *old man,* fallen under the awesome burden of sin, is to sin; then the nature of the *new man,* quickened by the power of God to a new life, is to glorify God and to enjoy Him forever!

Conceptual Thought

Conceptual Thought is information that lies at the root of all that God says and does. In it are His *personal thoughts*, a representation of the Omnipotent and Omniscient God. It contains the instructions for forming what amino acids and what proteins will produce the particular person He intended each of us to be; that will contain the blueprints for the construction of that unique person, none that will ever be identical to it. Each of us is *fearfully and wonderfully made.*

Information is non-materialistic. It does not contain carbon, phosphorus, or sugar molecules. Information is simply information, and no information can exist without a mental source in possession of a will. It is neither matter nor energy and is, by its very nature, a mental, and not a material quantity, a sort of mental telepathy system between transmitter and receiver; between God and man.

It was King David who wrote the most sublime Psalm in all of Scripture when he penned the 139th, in which he describes the unbroken and eternal relationship between man and his God. He begins from the point of not knowing or understanding all that there is to the creation of man; of what information and knowledge was made available; why and how was this wisdom applied; and finally ends up with revelations sublime for his time in history.

• **Search me and know me**
O LORD, thou hast searched me, and known me (Psa. 139:1).

What David did not realize was that God entered into a personal and intimate relationship with him, as He does with all His children; to reveal to David His deepest thoughts and the hidden secrets of His heart. He acknowledges that God has a perfect knowledge of him, and knows all his thoughts and desires. God searches out all those who love Him, penetrates the heart, examines intimately, knowing everything of those whom He created.

A penetrating understanding of being known of God is given in the book of Amos: *You only have I known of all the families of the earth* (Amos 3:2). Surely, God knows all the families of the earth, but He singles out the nation of Israel as He enters into a covenant relationship with His people. Of special interest to the Christian is that the Lord God has entered into a special relationship with him; searches him out completely; knows him personally; intimately; far above all life that God has created; whose highest thoughts are embedded into every nucleotide sequence in his body; animating his soul.

• Too wonderful, too high
Such knowledge is too wonderful for me; it is high, I cannot attain unto it (Psa. 139:6).

The knowledge of God is bewildering to me. Lord, I do not understand these things. Your thoughts surpass all my human comprehension and are past finding out:

> *Canst thou by searching find out God? canst thou find out the Almighty unto perfection? It is as high as heaven; what canst thou do? deeper than hell; what canst thou know?* (Job 11:7–8).

> I cannot grasp it. I can hardly endure to think of it. The theme overwhelms me. I an amazed and astounded at it. Such knowledge not only surpasses my comprehension, but even my imagination. . . . Mount as I may, this truth is too lofty for my mind. It seems to be always above me, even when I soar into the loftiest regions of spiritual thought. . . . We cannot reach to the lowest step of the throne of the Eternal.[1]

God is most knowable when we see Him incomprehensible; of how His thoughts fashioned us as He breathed in the breath of *lives*; and of how He upholds all things by the word of His power. He never loses anything, never forgets as we do, and observes and surveys perfectly the past, present, and future as if they were all there before Him in one instant. Not only are His activities as wide as the universe and as deep as the ocean, but His love for us is as high as heaven is from the earth. Infinity is no hindrance to Him. How amazingly inconceivable!

• Possessed and covered
For thou hast possessed my reins: thou hast covered me in my mother's womb (Psa. 139:13).

David realizes that God created or possessed him by *pro-curing him by purchase,* and that prior to covering or knitting him together in the womb. How much did David really know about the Biblical concept of purchase—the purchase and ownership of a sinner by blood—by the shed blood of Jesus Christ?

> *Before I formed thee in the belly, I knew thee; and before thou camest forth out of the womb I sanctified thee, and I ordained thee a prophet unto the nations* (Jer.1:5).

The Hebrew meaning of formed, *yatsar,* has several nuances, one of which is *to mould as a potter.* Before God the potter *covered* Jeremiah in his mother's womb, He determined who he was to be by *knowing* him, prior to his conception, determining what his life was to be like after birth. God knew what thoughts He was going to embed in his nucleotide sequence in order to function as prophet. He is involved in every person's life from the very beginning, not merely from conception, but from far back as eternity past.

• Fearfully and wonderfully made

I will praise thee; for I am fearfully and wonderfully made: marvelous are thy works; and that my soul knoweth right well (Psa. 139:14).

Who can gaze at the anatomy of the human body and still not marvel at its delicacy and precision? Do wonder and awe not swell up within you? Who will not pause and ponder the wonder of the event, to consider and tremble at the frailty of its stature? Is there not to be seen love and wisdom in its design, mercy and truth in its members, and worship and eternity in its soul?

We were known from eternity past and fashioned for eternity future. This present birth is but an infinitesimal part of our existence, waiting for the redemption of the body. What shall we say then of that new birth which is even more mysterious than the first, and exhibits even more the love and wisdom of the Lord?

David knew nothing about conception as we know it. He did not know that each of us starts as a single cell that comes from the union of a sperm and an egg. Nor did he realize that a cell divides into two, then four, then eight, then sixteen, on and on up to billions of cells. Nor did he know, or imagine in his wildest dreams, that at a certain stage in this process, one particular cell appears and that this one cell, as it continues to divide, ad infinitum, will become a brain, a heart, or a kidney—yea, it cries-out to be a person.

Who among us can adequately explain or elaborate on all the intricate workings of the human body. Do we *know right well* to offer praise and to marvel at the One who made us; not from flesh and bones per se, but from carbon, hydrogen, nitrogen, oxygen, and phosphorus atoms; at the One who formed us from the womb and to all the fashioning which precedes it.

What makes you and me so special to our God? What gives us so much value? Is it the right appearance, the clothes we wear, our position and title in the workforce? Might it be our bank account—social security, credit, or library cards? There must be something of worth that God foresaw in us that compelled Him to act in such a manner.

• Made in secret
My substance was not hid from thee, when I was made in secret, and curiously wrought in the lowest parts of the earth (Psa. 139:15).

In the council chambers of the Triune God, where all things came to be, the Word of God expressed what was intimately on His mind concerning the creation of man, and more specifically, the means, the method, and the time of his birth. Each individual element, each molecule, every nucleotide sequence, every amino acid codeword, and every protein in the construction of the body was laid bare before the Members of the council.

Who of us could have instructed Him as to the order and procedure of our making? Who among us could have informed Him as to the precise positioning of every element, of every organ, of every limb? But thank God I was made in secret, for *Who hath directed the Spirit of the LORD, or being his counselor hath taught him?* (Isaiah 40:13).

The Hebrew, *raqam*, translated *curiously wrought*, refers to embroidery, the intricate interweaving of various colored fabrics to produce a representation or picture upon the fabric, as recorded numerous times in the building of the tabernacle:

And the hanging for the gate of the court was needlework, of blue, and purple, and scarlet, and fine twined linen (Exodus 38:18), suggesting the beautiful artwork of the nucleus of the human cell, by which God expresses figurative messages, thoughts which are then ultimately read by the ribosome.

• Written in his book

Thine eyes did see my substance, yet being unperfect; and in thy book all my members were written, which in continuance were fashioned, when as yet there was none of them. (Psa. 139:16).

There is no unified consensus for this verse in the theological community for the first word in the Hebrew Bible is *glm*, translated as *my substance* or *my embryo*, and is the only time that it appears in the Bible[2]. But in the notes to this verse there is listed a *prp*, meaning, "it has been proposed" that another rendering be g*ml*, where the letters *l* and *m* are interposed. This indicates that *glm* is a possible defective word and the proposed rearranged rendering, *gml*, is also a possible rendering, and is to be translated as *my deeds* instead of *my substance*, or as *my embryo*.

Thus, is the verse referring to the fetus forming in the womb, or to the actions or deeds of an individual during a lifetime; to the formation of our bodies, or to works we perform? Laying aside textural criticism for another day, this verse is one of the strongest passages of Scripture confirming the doctrine of Predestination. God knows us intimately because He wrote it all down in His Book for you and for me to read.

• Your thoughts and me

How precious also are thy thoughts unto me, O God! how great is the sum of them! If I should count them, they are more in number than the sand (Psa. 139:17–18).

How precious are your thoughts of me, O Lord! Let me count the ways. Let me place them in terms of the DNA molecule and compare them to the sands of the sea.

The 46 spiraling DNA helical structures, the genome, contain information that is measured by its density; of how many nucleotide base pairs, or letters (like A with T), can fit into a cylinder transcribed by all the windings. This information packaging density is so inconceivably huge that it boggles the mind. As an example of the incomprehensible mind of God in conceptual thought, consider the following:

1. The total number of nucleotides per *haploid* genome = 3×10^9 letters
2. The length of the genome = 1020 mm
3. The volume of the genome = 3.19×10^{-9} mm^3
4. The volume of a pinhead of radius 1 mm = 4.19 mm^3
5. The number of genomes on a pinhead = 1.31×10^9

This indicates that there can be placed approximately 1.31 *billion* human genomes, in the volume of a pinhead! That is, the genomes of over 1.31 billion people; of who they are, male or female, of how they will control their lives, and of where, how, and when they will all live and die. All that information just in the volume of a pinhead! Six pinheads would adequately cover the world population of approximately 7 billion people! WOW! Now that's some genome!

But that's not all. Comparing the density of the DNA molecule to the density of a disc you have in your CD player, you would need to purchase 3.57 billion, 700 MB capacity discs, to equal the storage capacity of human genomes of 1.31 billion people located on the head of a pin. Pile-up the discs on top of each other and it will easily go around the circumference of the earth. Nothing short of astronomical! And at a cost of five cents per disc, you would have to cough-up 1.75 million dollars!

INFORMATION

There are instructions in the DNA molecule, information uniquely applicable for a particular protein. This could only have come into existence through intelligent, efficient design, not by random chance. The fact that it is found in every life-form is proof-positive that a single Designer created all life. Encoded messages can no more be attributed to nature than could the writings of Shakespeare or to the music of Bach. To create great works of literature requires an intelligent life; to create life itself requires an omniscient God.

No naturally occurring molecule possesses the properties of information. Nature does not produce any kind of code or an encoding/decoding mechanism. Everything in nature, other than DNA, represents only itself. There is no known mechanism by which natural processes produce information. DNA on the other hand, represents a complete plan for a living organism, unique to that individual life; an encoding/decoding machine; a language that uniquely represents the organism.

THE DIGITAL COMPUTER

By analogy, the human body can be compared to the modern computer where both entities seem to function under the influence of codes peculiar to that entity. What is truly revolutionary about molecular biology is that it has become digital, which has the added benefit that this type of code may contain gigabits of information as compared to its analog sibling.

The information contained in the digital bit string "1001 0110," one byte, means something to a computer, put there by the designer. The designer (the transmitter) knows what it means for he has a specific mental thought to transmit to the computer (the receiver) for processing. He knows what it means and how the computer will react to the code. It could possibly mean to open up Microsoft Excel or Word, to download a file from the Internet, or to inform you that: "You have mail!" One bit out of sequence will give you that dreaded error message: "Your computer will shut down in five seconds!"

THE DNA COMPUTER

The digital nucleotide sequence ATC GGA TCC, three consecutive 3-bit codons, means something to a human cell, put there by the Designer. He knows what it means and how the cell will ultimately react to the code. The code could mean that you are "beautiful and blond," or "handsome and six feet tall," or perhaps "$x^2 + y^2 = r^2$," or maybe it could possibly allude to the finale of the "Hallelujah Chorus." The code will mean *exactly* what the designer intended it to mean, who, by the way, is remote or external from the computer.

What all this means is that the DNA in every life carries an encoded message unique to that individual, digitally embedded in the nucleotide sequence of DNA. It is the message of life spoken in every living creature on the planet, and from time immemorial, the questions that echo in the hearts of men throughout the entire universe are: *What message? A message from where? A message from whom?*

sDNA is a fantastic molecule, beyond our capacity to describe it. How is the message coded that spells out the recipe that makes an arm instead of a leg, or a heart instead of a kidney? What does it *ride* on? Some may think that DNA is alive, this is not so. DNA of itself is a dead, lifeless molecule. Sugar, phosphorus, and bases are not alive. It may seem as if DNA is the actual information in your body. Not so, DNA is simply the medium on which the message is written and stored. Essentially, the DNA sequence performs the function of an encoder, rewriting the message into suitable formats to be processed further, and then decoded when called upon to do so.

Let us go back once again to the first creation of Adam when God formed man of the dust of the ground, and let's halt the process at the end of this first step. Adam can be seen as lying on his back on the ground, not breathing or moving, seeing, or speaking. Adam is dead as we know it with no life principle within his body. We can draw the conclusion that Adam's nDNA body is incomplete, in need of more specific inputs from his Maker in order to become a living soul. His DNA is dead!

Genesis 2:7 continues, illustrating the second step in creation: And God *breathed into his nostrils the breath of life.* Whatever you wish to assign to this statement, one thing is sure; life at its basic core is information. No one moves or breathes without information, saturated and embedded in all his members. All information pertaining to our bodily functions are written into our genes via the "breath of life" of God.

First, there is a flow of information from an external source to the non-living DNA molecule and embeds within the nucleotide sequence of the molecule. The DNA molecule knows enough to make a replication of itself that will be used to form a template, off of which specific genes are copied to mRNA in transcription; the message contained in DNA is encoded in mRNA. Nothing has been lost in transcription; information has been added to the original message to aid in error detecting and repair. Translation is decoding the encoded message in mRNA in order to produce the polypeptide chain of amino acids that, when folded upon itself, will form a 3-dimensional molecule. That protein molecule is the particular molecule contained in the input information source to the DNA. Once the protein has been assembled and folded, it can then go to work to form its intended biological function.

SPECIFICITY

To form a protein, amino acids must link up to form a continuous linear chain, at times containing up to hundreds of amino acids. This chain must be highly specific in its sequence in order to be a biological, functioning protein. Amino acids unspecified do not make a functioning protein any more than mere letters make words or sentences of poetry.

Specific one-dimensional amino acid sequences are mapped into specific, very precise, three-dimensional, folded proteins. This structural shape, a twisting, turning, tangled chain of amino acids, in turn determines what function the amino acid chain can perform within the cell. No three-dimensional shape, no functioning protein. A one-dimensional shape cannot function as a biological protein, but a three-dimensional shape gives it a "hand-in-glove" fit with other molecules in the cell.

Because of specificity, one protein can usually no more substitute for another than a screwdriver can substitute for a table saw. It will only fit with other equally specific and complex three-dimensional molecules within the cell. Proteins may travel throughout the body in the bloodstream and will bond to other molecules, but they will not bind to molecules that do not exhibit the very same specificity as themselves.

THE WORD AND THE LIGHT

The genetic code embedded in the DNA molecule is akin to a book that contains letters, words, sentences, and paragraphs. It is a language not unlike the language of English or Chinese; the language of the stars or the language of the planets; the language of mating calls of butterflies or sparrows; or the language of computer machines and iPads. It is the universal language of life, a language similar to what you are now reading on this page; and that it can only come from a mind that designs, thinks, and speaks *words* in order to build complex systems; possessing a mind that is highly complex, yea, Omniscient!—and His name is:

THE WORD!
And he was clothed with a vesture dipped in blood: and his name is called The Word of God (Rev. 19:13).

His name is *The Word,* a name no one knows but He Himself; only this we know, that this Word is Jesus, made manifest in the flesh. The Word (Gk, *Logos)* is described as:

Something *said* (including the thought); by implication a *topic* (subject of discourse), also *reasoning* (the mental faculty) or *motive;* by extension a *computation;* specially (with the article in John) the Divine Expression (i.e. Christ): ---communication, say(-ing), shew, speaker, speech, talk.[3]

The Greek *logos* means *a thought or concept* and then followed by *the expression or utterance of that thought.* As a designation of Christ, logos is peculiarly felicitous because in Him are embodied all the treasures of divine wisdom, the collective thoughts of God; and He is, from eternity, the utterance or expression of the thoughts of Deity. Adam Clarke describes the Word as the:

Logos, which signifies a word spoken, speech, eloquence, doctrine, reason, or the faculty of reasoning, is very properly applied to him, who is the true light which lighteth every man who cometh into the world; who is the fountain of all wisdom; who giveth being, life, light, knowledge, and reason, to all men; who is the grand Source of revelation, who has declared God unto mankind; who spake by the prophets; who has illustrated life and immortality by his Gospel; and who has fully made manifest the deep mysteries which lay hidden in the bosom of the invisible God from all eternity.[4]

How appropriate is the comparison of the DNA sequence to a spoken language, to realize that the name of the creator of every living being in the universe is the Word. In the Person and work of Christ, Deity is expressed:

The Scriptures reveal God's mind, express His will, make known His perfections, and lay bare His heart: (a) a "word" is a *medium of manifestation.* I have in my mind a thought, but others know not its nature. But the moment I clothe that thought in words it becomes cognizable. Words, then, make objective unseen thoughts. This is precisely what the Lord Jesus has done. As the Word, Christ has made manifest the invisible God; (b) a "word" is a *means of communication.* By means of words I transmit information to others. By words I express myself, make known my will, and impart knowledge. So Christ, as the Word, is the Divine Transmitter, communicating to us the life and love of God; (c) a "word" is a *method of revelation.* By his words a speaker exhibits both his intellectual caliber and his moral character. By our words we shall be justified, and by our words we shall be condemned.[5]

In the beginning was the Word, and the Word was with God, and the Word was God. . . . All things were made by him; and without him was not any thing made that was made (John 1:1, 3).

You cannot come into the world without the Word and without the Light. They are the instructions that are passed on to human cells, to make proteins that go to make body organs that identify each person from every other person.

THE LIGHT
The spirit of man is the candle of the LORD, searching all the inward parts of the belly (Prov. 20:27).

Strong defines *light* (Gk. *phos*) as meaning *fire* or *light.* Also pertinent to our discussion is *phemi*, which has the same root as *phos* and has the meaning of *to make known one's thoughts, to declare or to say.* Both words come from the obsolete verb *phao* that has the meaning of *to shine or make manifest by rays.*[6]

We may draw two conclusions from the above definitions: 1) that *light* is something said and comes from a mind that has thoughts that are to be declared, and 2) that this leads us into the realm of *photons* that we find in the realm of the electromagnetic radiation spectrum, information transmission *by rays* of energy. Light is *photons,* the carrier of information. Information modulates the carrier and the carrier is the medium the information rides upon, the conceptual thoughts that animate the message. It is the Light that permeates the DNA molecule, biologically demodulated, and then embeds itself within the re-arranged nucleotide sequence that enlightens the body of a Christian. It is the Light that shines throughout the world of sin and darkness to give the gift of life:

For God, who hath commanded the light to shine out of darkness, hath shined in our hearts, to give the light of the knowledge of the glory of God in the face of Jesus Christ (2 Cor. 4:6).

Light comes to us in two different flavors. At times it acts as distinct particles and at other times it behaves as a wave function. Which one is active at any given moment or circumstance is hard to visualize. A Conceptual Thought process emanating from the throne of the Word of God, if in the form of particles, may enter the body at specific locations; typically at the frontal cortex of the brain; causing disturbances that excite the hypothalamus, amygdala, and the pituitary gland to secrete specific neuropeptides, regulated by the message contained in that particular thought; to then travel along hard-wired neuronal pathways to other parts and organs of the body that the pituitary dictates. Or, if the thought comes to us as a wave function, then the entire body is bathed in the Light, as specific neuropeptides race up and down fluidic passageways of the body; flooding every organ and system, seeking receptors with the same specificity that connect with it; exciting the cell to function as directed.

The Light in the gospel of John is not the same light as the light in the book of Genesis. The light in Genesis shines, targets, and embeds in natural nDNA; whereas the light in John shines upon all living men that put their faith in Him for salvation, targets, and embeds in spiritual sDNA. The light of Genesis contains no informational content as pertaining to eternal life, whereas the Light of John contains it all.

I am the light of the world: he that followeth me shall not walk in darkness, but shall have the light of life (John 8:12).

CHAPTER THOUGHTS

The informational density in a DNA molecule is staggering. It contains the highest information density ever known to man. There is sufficient capacity in the DNA molecule for God to insert, as Conceptual Thought, all that is needed to be there to spell out the life of a person. He has included everything that is possible to occur; therefore every aspect of our lives must be included in the code if it is ever to be realized. Everything is there that is possible to be, but nothing is there of which it is impossible to be. We will never know the myriad contents of our lives by just observing the code, trying to decipher its meaning. Whatever we do, whatever we think, whatever we say or read, all the activities we do in life come directly from the messages embedded in the nucleotide sequences impressed upon with the Word of God. All human intelligence is housed there, waiting to be revealed.

Consult the great scientific minds if you will, search all the books in all the libraries in the world for the meaning of the code, and you will not succeed, for you are attempting to intimately know the mind of God. How divine thought, impressed upon non-living chemical nucleotides to cause life to be played-out in the human body, is an eternal mystery and is the Holy Grail of biology. How then can we talk about that which cannot be defined or analyzed, and which is more likely to elude us the harder we search for it?

Are we about to be caught in a dilemma similar to that of the poor soul who once asked Louie Armstrong what Jazz is, only to be told: "Man, if you gotta ask, you'll never know," seems to be the dilemma anytime we undertake to explore spirituality in a verbal context.

Without a doubt, the most complex information processing system in existence on earth is the human body.

And so it is with the new Christian.

17

LONGSUFFERING, GENTLENESS & GOODNESS

*"But the fruit of the Spirit is....longsuffering,
gentleness, goodness...."* (Galatians 5:22)

We have discussed what we believe as to the method God might have used to regenerate an elect person. In previous chapters, we have seen how the virus selects a cell in the body, mounts an attack, and penetrates the cell membrane to gain entrance into the cytoplasm. Then, once entrance is gained into the nucleus through nuclear pores, changes in the DNA structure then codes for more viruses of the same kind; then emerges out of the cell to roam throughout the body to present itself to other cells. These changes in DNA structure may be advantageous in the cloning of beneficial viruses that act as antagonists, or detrimental in order to spread its venom throughout the body.

What God did with the virus, He can also do with us. In like manner, upon the initiation of regeneration, once the Message of God gains entrance into the nucleus of a degenerate cell, the host nDNA is infected, then metamorphosed into spiritual sDNA. The Holy Spirit immediately administers the full benefits of the *fruit* within the regenerated cell; spiritual sLove and its siblings run rampant. The Message then exits the cell, budding-out from the cell through the plasma membrane into the intercellular fluidic pathways; and then roams throughout the body in search of nDNA flesh marked-out for salvation, *to seek and to save that which was lost* (Luke 19:10).

THE LIMBIC SYSTEM

We as humans will never lose our humanity. God works with what we got. Our spiritual cells look no different under an electron microscope than do our natural cells. He uses the analogy of the common virus to explain, in ordinary scientific terms, just how regeneration and sanctification may have been accomplished.

At that moment in time, by means of His sovereign grace, God the Father ordained regeneration to occur in each of us; dispensing a source of divine energy capable of performing what He has ordained from eternity past. He *infects* the human body by seeking out cells that contain specific cellular receptors that are identical to, and have an affinity for, spiritual life. This action by the Father is unbeknownst to the recipient but issues in the new life in Christ.

The message of eternal life, now *resident* in neuropeptides, bonds onto surface receptors that He has made willing, and immediately the divine information is transferred into the cell cytoplasm, then on into the nucleus and ultimately into the nucleolus. Jesus Christ places into effect the essential programs for cellular DNA to function in a spiritual atmosphere; and the Holy Spirit proceeds to act upon the newly embedded information by transforming natural flesh to spiritual sDNA flesh. Regeneration of infected cells is now made complete.

THE EMOTIONAL BRAIN

The limbic system appears to be the part of the brain responsible for translating thoughts and ideas into emotions and feelings. The attributes of the *fruit* would be transmitted by the Spirit to the frontal cortex that acts as a receptor area to deal with signals associated with thought; where incoming messages would be evaluated, stored, and retransmitted to the limbic system; the main components consisting of the hypothalamus, amygdala, pituitary, and the hippocampus, which governs the pituitary gland.

The divine attributes of the fruit of the Spirit, when embedded in the mind of a new Christian upon regeneration, produce qualities in the body that are spiritual, and occur primarily within the realm of our emotions. To comprehend the magnitude of this event, we need to affirm that we are but a small analogy of the divine nature of God and that we have emotions because God is emotional. We are in the image of God's emotional image, and when we speak of God having emotions, we can "feel" those emotions, albeit to a lesser degree, in our spiritual emotions.

Spiritual peptides and immunes are similar to the natural but function differently. Whereas the natural allows both good and bad things to occur in the body, the spiritual allows only the good. The natural peptide oxytocin may produce the emotion of natural love, but it also may not hinder the emotion of hatred; whereas spiritual oxytocin will only produce spiritual love continually, locking out hatred.

As the Holy Spirit indwells the Christian believer, He assumes control of the body, for His thoughts now become, to a limited degree, our thoughts. Whenever a neuropeptide binds to a receptor, the receptor takes on, in its memory, the content of the chemical information associated with the peptide. Therefore, when the Spirit injects thoughts of Love, Joy, and Peace into receptors with the same specificity, this information is constantly made available to maintain a level of psychological and physiological "feelings" of love, joy, and peace, swaying harmoniously in synchronism throughout the body:

> The Holy Spirit—*the conductor*—orchestrating the symphony of life and health; the peptides—*the sheet music*—containing the notes and rhythms that allow the orchestra—*the body*—to play harmoniously in unison with the mind; the receptors—*the players*—swaying majestically and melodiously in tune with the musical feelings of emotion and pleasure—all emanating from the fruit of the Spirit.

Both our spiritual body and our natural body enjoy the blessings bestowed by the *fruit*. As God the Father finds a biological similarity in viruses that bud-out from the nucleus of virally infected cells in order to invade other cells for regeneration, in like manner, the fruit of the Spirit buds-out from every regenerated cell, ready to invade every newly regenerated natural cell with the Love, Joy, and Peace attributes of God; spreading spiritual love, joy, and peace in the hearts of Christians for the things of Christ.

SPIRITUAL FRUIT: GROUP 2

The second group of three fruit of the Spirit relates to our responsibility toward others, of how we are to live and act, not only toward believers in Christ, but also toward unbelievers:

- **LONGSUFFERING → sGABA → sLongsuffering →
    sT Cell**

But thou, O Lord, art a God full of compassion, and gracious, longsuffering, and plenteous in mercy and truth (Psa. 86:15).

THE LONGSUFFERING OF GOD

The Longsuffering (Gk. *makrothumia*) of God is patient with man despite his rebellion, for He is slow to anger and abounding in mercy. No greater demonstration of longsuffering can be found than that shown by God toward those who place their faith in His Son Jesus.

• sGABA

It is said that GABA (gama-AminoButyric Acid) inhibits (relaxes), hence the *"inhibitor peptide,"* while glutamate excites (anxiety); proportionally secreted together, they keep the body in an emotional balance. GABA applies the "brakes" on emotional excitability, calming down neurons, keeping them at rest; while glutamate steps on the "gas," gets you going. With too much GABA—brain activity shuts down, stopping vital functions; too much glutamate—seizures.

Serotonin, the joy peptide, seems to be the mediator in establishing the optimum secretion levels of GAMA and glutamate to produce an emotionally spiritual, longsuffering attitude—of patience and forbearance in the midst of adversity. Is it not surprising then that each fruit of the Spirit bears a direct relationship to all the other fruit.

• sLongsuffering

The fruit of spiritual longsuffering follows immediately after love, joy, and peace, and deserves the right to occupy this coveted position. Throughout Scripture, longsuffering and its sibling, patience, is most often spoken of than all the other fruit that follow. Why? Because there is power in spiritual longsuffering, for it is Jesus who clearly epitomizes longsuffering and patience. Who can read the amazing story in 1 Peter 2 and walk away without being moved with His display of long-suffering, clearly holding Jesus up to us as *the* example we ought to emulate:

For what glory is it, if, when ye be buffeted for your faults, ye shall take it patiently? but if, when ye do well, and suffer for it, ye take it patiently, this is acceptable with God (v. 20).

Be like Jesus! When, for all your well doing, ye suffer in-dignation for it—take it patiently, leave an example, sin not, have no guile, do not revile, and threaten not—*this is accepta-ble with God*. It is often quoted that one that is longsuffering, suffers long. Natural acts of "the works of the flesh"—*nUncleanness*—the antithesis of sLongsuffering, the first work in an unbeliever's arsenal of attitudes directed toward his god, the Antichrist, now presents no formidable foe to the believer and are, therefore, permanently eradicated.

• sT Cell
T Cells find and kill but they do not *eat*. As with their cousin, the Natural Killer (NK) cells, these cells are the ulti-mate killing machines! T cells are very intelligent. They think, plan, cooperate, and execute. They communicate with each other as no other army on earth can do. They are fast. They react in nanoseconds to an attack. It is amazing how they can recognize and differentiate between thousands of different molecules. T cells, in harmony with the neuropeptide GABA, produce a defense system unequalled in performance, afford-ing a superior measure of spiritual longsuffering.

• GENTLENESS → sDopamine → sGentleness → sB Cell

Thou hast also given me the shield of thy salvation: and thy gentleness hath made me great (2 Sam. 22:36).

THE GENTLENESS OF GOD
We can look to the best example we have of a person who modeled the Gentleness (Gk. *chrestotes*) of God. You guessed it; it was Jesus. The apostle Paul exclaimed: *Now I Paul myself*

beseech you by the meekness and gentleness of Christ (2 Corinthians 10:1); gentle as a lamb when being led to the slaughter!

• sDopamine

Dopamine, *"the pleasure peptide,"* affects diverse areas such as mood and appetite and thus is involved in reward and reinforcement. It seems that every time we experience fun and enjoyment, or we do things that are pleasurable and interesting, the body releases dopamine to sustain these activities. This includes listening to enjoyable music or eating chocolate. Dopamine insures that we will, at some later time, repeat the same behavior over again.

• sGentleness

But we were gentle among you, even as a nurse cherisheth her children (1 Th. 2:7).

When I think of gentleness, I think of holding a brand new baby. Speaking a soft, tender word, especially to a new child, turns aside bad feelings. The kind of gentleness God is talking about is similar. And, you know what I've learned?—being harsh towards someone else never solves a problem. It just engenders bad feelings, for *a soft answer turneth away wrath: but grievous words stir up anger* (Proverbs 15:1).

Spiritual gentleness is a trait to have when dealing with people. To be gentle is to be considerate, amiable and tender. Wow! I never considered that being gentle is to be tender. How often do we hear that word? And how often do we treat other people with love and tenderness? What is this really saying? Are we trying to say that God is gentle and— *tender?*

Certainly Jesus was. I am learning that God values gentleness, and yes—tenderness—in our dealings with people for gentleness is not being weak but the opposite is true— gentleness is strength, which allows us to be tender. He who transcends space and time, He who is bigger than the universe itself, vast and powerful, yet He extends to us His utmost in tenderness and care.

Gentleness alleviates natural diseases of—*nSeditions*—the antithesis for sGentleness, stemming from the "works of the flesh," the second work in the war against unbelievers aimed at pleasing the Antichrist. Satan is becoming totally defeated as more body cells are coming to believe in the Lord.

• sB Cell

By this time in your reading, has it yet occurred to you that everything is composed of amino acids? DNA, proteins, immunes, immunoglobulin, and a host of other particles roaming throughout the body are all amino acids in one configuration or another. Let's face it. We can't get away from it. We are all basically aminos!

B cells make immunoglobulin that bind to pathogens via attached antigens to enable their destruction; but after an attack, they retain the ability to produce memory antibodies; cells that will never lose their memory. sB cells can remember every sin antigen/pathogen complex attempting to gain admittance to the body and immediately set out to destroy the intruders. They are always successful.

• GOODNESS → sEpinephrine → sGoodness → sNK Cell

And he said, I beseech thee, shew me thy glory. And he said, I will make all my goodness pass before thee (Ex. 33:18–19).

THE GOODNESS OF GOD

Moses asked to see the glory of God and was blessed to view all of His Goodness (Gk. *agathosume*). God is the source of everything that is good and epitomizes the sum total of all His attributes. God alone is Good. You cannot have goodness without God and you cannot have God without goodness. God's goodness is absolute. All others have degrees of goodness as measured against this absolute standard. It permeates all space and is incorporated into everything He is and does. His holiness, His righteousness, and even His wrath are all good. There is nothing intrinsic in God that is not good, and there is nothing in God that He wills for His children that is not good, yea, perfect.

• sEpinephrine

Have you thought much lately about laughter? Did you know that it is good to laugh! Did you know that children laugh about 400 times a day, but adults only fifteen! Listen to your children! They would be quick to tell you that a good belly laugh is good for your health. God made it that way. It enhances blood oxygen, increases circulation, stimulates alertness, and reduces inflammation. Laughter is aerobic, an inner jogging, akin to physical exercise in which all the major systems of the body are exercised.

Now everyone knows that the hypothalamus-pituitary-adrenal (HPA) axis is involved when laughter is expressed. The hypothalamus gets excited and secrets the hormone CRH which finds its way to the pituitary gland. The pituitary then secretes the hormone ACTH that finds its way to the adrenal glands that sit atop the kidneys. From there, epinephrine, alias adrenaline, is released to flow throughout the body, and if released in the proper amount, has anti-inflammatory properties that reduce stress and depression by enhancing blood oxygen; increasing circulation, reducing inflammation, and stimulating alertness—hence the designation *"the fight-or-flight peptide."*

The release of ACTH out of the pituitary produces epinephrine downstream that makes one feel very good, and in turn, encourages one to feel like laughing more. The circle is unbroken; laughter is in a feedback loop back to the pituitary that now starts to produce more epinephrine; so the more you laugh, the more you want to laugh; and the more you want to laugh, the more you produce epinephrine. What a system! You know, sometimes I just can't stop laughing!

• sGoodness

God is the source of everything that is good. To be near to God, to have intimate fellowship with Him, should be the aim of every Christian. Whatever happens to us in life, whether we deem it to be good and of the will of God, or whether we say it is bad and God is punishing us, whatever we say or think, God meant it for our good: *But as for you, ye thought evil against me; but God meant it unto good* (Genesis 50:20). When God brings suffering into the lives of His people, as He at times chastens His children, goading us to remain near to Him, we should never stray afar as He will ultimately reassure us that His goodness will always remain.

The good man is one who thinks about love, beauty, and truth—not just in the realm of majestic mountains, surging seas, gorgeous flowers or beautiful sunsets, but also specifically in his fellow man. His wish is to alleviate suffering and to mitigate wrongs. He consciously looks for ways to benefit others. His works are the opposite of the self-centered works of darkness. To be good as God is good is to eliminate natural—*nWitchcraft*—the antithesis of sGoodness, the third work in the "works of the flesh" that honors the Antichrist.

• sNK Cell
The art of medicine consists of keeping the patient amused while nature heals the disease (Voltaire).

Natural Killer cells are yet another type of lethal lymphocytes containing granules filled with potent chemicals. They are called "natural killers" because they do not need to recognize a specific antigen before swinging into action. They do not attack invading organisms directly but instead, they destroy the body's own cells that have become infected.

NK cells are the "marines" of our white blood cell forces. They kill on contact. The killer binds to its target, aims its weapons, and then delivers a lethal burst of chemicals that produces holes in the target cell's membrane. Fluids seep in and leak out, and the cell bursts. Think Rambo!

There is no greater debilitating stress to the body than that of hearing the dreadful words, *"you have cancer!"* Those words cause chronic stress that exerts a suppressive effect on the immune system of the body. Anything that causes a decrease in immune system function must be eliminated ere the disease goes uninhibited.

Enter once again—laughter! Laughter may reduce stress by increasing NK cells, interleukins, and interferon, thereby improving immune cell activity. Immunoglobulin A (IgA) helps protect against upper respiratory infections. Scientists have documented significant increases in IgA in response to comedy programs designed to produce a great deal of laughter. They recently have claimed that watching a one-hour musical or comedy program could increase the number of B cells, T cells, NK cells, and gamma Interferon in our body; all the good guys! There is something about humor and laughter that causes the immune system to turn on. True laughter does not hurt, is not deceitful, or harmful, and it always feels so good to laugh!

CHAPTER THOUGHTS
We have continually stressed the Biblical teaching that the nNature governs the natural flesh, and the sNature governs

the spiritual flesh. These two natures with their divisions of flesh are at war with each other. nDNA flesh is at first more abundant in quantity of natural cells and therefore more powerful than sDNA cells. At this stage, the battle is in favor of the natural man who exhibits the "works of the flesh" throughout his life, while the spiritual man struggles to produce the moral values of the "fruit of the Spirit."

The spiritual man continually seeks to obey the commandment to "put on Christ," to put on His divine attributes albeit to a finite degree. God Himself produces the fruit that the Christian may grow up into the full stature of Christ. Each of the Members of the Godhead participates in this endeavor. And guess what—He uses ordinary bodily functions to accomplish His work, starting at regeneration and up to sanctification. Moreover, He starts down at the molecular level.

When specific receptors mate with either the same specific neuropeptides or immune components, the receptors vibrate or wiggle (does a dance), and certain physiological and emotional changes happen throughout the body (now if that isn't awesome, I don't know what is). Now add to these strange phenomena, viruses. This means that viruses vie for the very same receptors as does neuropeptides and immunes! This insures that our three major systems of the body are intimately related and physically connected. When one acts, the others know full well the ensuing results. It is a tough call, but the body makes these decisions millions of times every second!

It may be that God, acting now as would the virus when exiting from regenerated cells, simply enters other marked receptors and institutes regeneration directly and then injects fruit or immunes. Perhaps, prior to this action, He would allow the fruit to seek out and mate to specific receptors to spread the Love of God into the cerebral cortex to remind the as-yet unregenerate elect of their sin and the need for regeneration. Or, perhaps, He would allow immune cells to first mate with receptors to reward the recipient with good health in order to show forth the Goodness of God by paving the way for the administration of the *fruit*. Whatever the method God uses, it *feels so good to be Christian!*

18

THE NEW BIRTH EMBRYO:
THE YOUNG MEN

*"I write unto you, young men, because ye have
overcome the wicked one"* (1 John 2:13)

Everything starts with a birth, both your natural life and
your spiritual life. You and I started our spiritual lives
after having been called-out by the Holy Spirit. The
process is spiritual; no one can adequately explain it. You may
quote the scriptures to build your faith, but the process of how
this is done is supernatural and will forever remain a mystery.

Do you remember when you were first born? Do you remember how all that happened? After all, you were just a baby.
The experience is the same when you were born the second
time. You could not understand what was going on in your life
at that moment, what it meant to be saved, or what lies ahead.
It was simply impossible for you to know those things at this
stage in your spiritual life. You needed some time to be nourished, some time to grow up, some time to gain experiences.

The Bible offers spiritual milk for spiritual babes. Fundamental drinking instructions are the first step to start new
lives. Your pastors and teachers act as spiritual mothers, all
you need to do is listen and then drink. However, milk is sufficient food only while we are babes. We will soon need other
nutrients to keep growing. Drinking milk helps baby's bones
grow stronger so that we can start crawling, then standing,
and finally begin to walk and to talk. It is the fundamental
food, but as the development stages change to higher levels, so
must our food diet change.

SPIRITUAL CHILDBIRTH: PHASE 3

JUSTIFICATION

Following the act of regeneration and reconciliation, the Holy Spirit institutes the act of *Justification* to continue the birth process. Justification produces a spiritual embryo in the body, i.e., *Justification → Organismal → Embryo → Young Men*, continuing the process of the spiritual birth of a son of God. The new birth embryo contains all the genetic characteristics of its parents. It is neither the mother nor the father of its natural parents, but is characteristic of its spiritual parents.

Justification is God's act of declaring or pronouncing a sinner righteous, and therefore sinless before the throne of God. The Reverend Billy Graham always referred to the act of justification as: *"Just as if I'd never sinned."* Justification has the connotation of—"to declare or to make one righteous" in the sight of God, presented as a satisfaction for God's wrath upon the sins of mankind. It entails the exchange of imputed righteousness for sin—the sacrificial works that Jesus accomplished in His life are imputed to His people, while their sins are imputed to Him at the cross. The Lord shall finally *see his seed,* and *shall be satisfied.*

> *He shall see his seed, he shall prolong his days, and the pleasure of the LORD shall prosper in his hand. He shall see of the travail of his soul, and shall be satisfied* (Isa. 53:10–11).

ORGANISMAL

As in the analogy of the good tree planted in good ground in the physical realm, and as yielding good fruit in the spiritual realm, so also the embryo begins to grow in grace to ultimately produce spiritual young men. As there must be an inward growth in the tree, there must also be an inward growth in the young man; though exactly what it is or how it is accomplished is not fully known:

> Is not the inward growth of a Christian that aspect of his progress which is the most difficult to define, describe, and still more so to put into practice? Unless the tree grows inwardly it would not grow in any other direction,

for its outward growth is but the development and manifestation of its vital or seminal principle.[1]

The inward growth of a tree consists in the toughening of
its tissues or strengthening of its fibers; the internals of an
older tree is harder and stronger than that from its sapling.
The spiritual counterpart is found in obtaining more strength
of character, so that the Christian is no longer swayed to-and-
fro by varying opinions; becoming more stable, less emotional,
and more rational; wiser in spiritual matters; increasingly en-
gaged with the Word of God; more settled in doctrine; and
more discriminating in what he reads, hears, or says.

EMBRYO

Justification is the signal that opens the door for the blastocyst to divide into a larger number of sCells to reach the status of embryo. The embryo invades and embeds in the
endometrial lining of each natural organ in the body. *Villi* invade each placenta to establish nutrient pathways from each
natural host to the new embryo. The new birth embryo is the
sum total of all the blastocysts attached to each host organ
system in the body. In this manner, every sDNA molecule attached to every organ of the natural host contains, in some degree, the image of God.

But, before embryos can survive, either natural or spiritual,
and sons and daughters be born, they have to survive a potentially fatal conflict with their mothers. The embryo of a Christian is essentially a foreign tissue within the womb and is
subject to triggering a hostile maternal immune attack against
the invader. *This is essentially an invasion upon the natural
flesh of the host by the spiritual flesh of the new embryo.*

Before birth, the spiritual embryo's immune system is not
fully developed and, therefore, cannot of itself produce most of
the essential antibodies required for its protection against the
embattling immunes of the host. Therefore, most of the antibodies needed by the embryo must be acquired from the host
directly. The new placenta counteracts the natural tendency of
the host organ to reject the foreign tissue of the embryo,
thereby enabling the transport of antibodies from the host to
the embryo—*passive immunity*—the passive acceptance of the
host's antibodies. In this wise, the placenta forms an immunological barrier across which sin antigens from the host organs

dare not cross, actively shutting down the host's natural defenses with respect to the embryo.

Nor can blood from each host ever cross the umbilical cord
into the embryo, causing the sin antigens of the host to invade
and contaminate the embryo. The new birth embryo, therefore,
is forever settled and secured. You can never lose your new
birth biological status, which is akin to never losing your theological salvation.

Spiritual realities slowly begin to emerge and can be clearly discerned as the Holy Spirit gives utterance. We as embryos
need to start putting away some of those filthy things in life
that we once loved and held so dear to our hearts; cursing,
fighting, frequent trips to the tavern, cheating on our spouses.
But be very clear about this; this is not an easy task to accomplish for a person who, all their lives, lived in the depths of sin
and despair, not knowing that there was anything morally
wrong. It is not you personally who can accomplish this by
your own power. What little progress is made at the beginning,
you will begin to wonder if you are still in your sins, on your
way to a Christ-less eternity in hell. Be of good courage, for
the Holy Spirit has not left you desolate. You may falter or
stumble but you will never lose your position.

The battle, begun at conception, passing successfully
through the zygote and blastocyst stages to reach embryo, still
rages and is just now heating up. As the caterpillar enters the
cocoon and begins the process of metamorphosis resulting in a
new creature, cell by cell, into a beautiful butterfly, so also the
new man begins his journey; growing and developing, cell by
cell, into a beautiful person endowed once again with the
tselem image of God in every cell. The process of growth and
resurrection has begun and is ongoing; both take time and effort to complete, but complete they will.

Embryos need constant care and the proper nutrients to
grow. God ceaselessly works through his grace to impart to,
and perfect in him, all that He desires him to be. But as time
goes on, as spiritual growth happens, the embryo, like the butterfly emerging from the cocoon, begins to see the light of day.
The embryo begins to think and know of spiritual things, and
to feel and react to the outside world, more so than the zygote
or blastocyst.

Genetic traits exist at the organismal level. The red color
in the bloom of a flower is a genetic trait. A plant has an

enormous phenotype that includes everything from its shape, its leaves, its flowers, its cells, its enzymes, and even the color of its bloom. The Christian grows from the moment of conception to the time of rebirth, mimicking the growth of a simple red rose. From the time he was called-out by the Holy Spirit, passing through regeneration at the molecular level to initiate conception of the zygote; receiving reconciliation at the cellular level to advance to a blastocyst; and now justification at the organismal level; each level clearly exhibiting visible marks of Christian growth.

YOUNG MEN

The embryo is no longer a babe or little child of God, slow in growth, barely able to receive the milk of the word, an immature Christian, but is now referred to as a *young man* in Christ.

There is a great deal of growth in the embryo from that which is said of babes and of little children: Babes are said that their *sins are forgiven,* and of little children, *ye have known the father.* These distinctions are accumulative as time goes on throughout the birth process and build upon successive marks associated with a Christian. There are four states in the birth process, and there are particular duties associated with each state. As young men in Christ, Scripture reveals several patterns of behavior, images of Christ that can be claimed by all Christians at the embryonic level. The apostle John writes words of encouragement to the young men and proclaims that:

• **Ye are Strong**
I have written unto you, young men, because ye are strong (1 John 2:14).

Weakness is a sure sign of an immature Christian, but strength is a blessing from the Lord. Weakness is made manifest in natural nDNA flesh but strength comes from spiritual sDNA flesh. *Ye are strong* because you regained the image of God in your genome, the image that was lost when you first sinned, but is now regained. It is not the spiritual aspect of the young men that must be written and preached unto, as the spiritual is in need of no instruction, but to the natural. In this respect, none of us are too old or too young to learn.

Many young men are strong in faith but some are still weak in development. Young men are renowned for their athletic vigor and are the ones called upon to fight in the defense of their country. Proverbs 20:29 tells us that *the glory of young men is their strength,* and are pictured as victorious in conflict, for they are better versed in the Word of God.

• Ye have the Word
I have written unto you, young men, because . . . the word of God abideth in you (1 John 2:14).

This is the secret of strength and the source of victory— *the word of God abideth in you.* It is of a truth that all the great Christians have been men of the Word. In fact, John says you're strong because the Word abides in you and, because the Word abides in you, ye have overcome the wicked one. The secret of young men's strength is the Word!

Our Lord relied upon the very same principle. When He was tempted by Satan in three different locations, in the wilderness, on the mountain top, and in Jerusalem, three times He said: *"it is written... it is written... it is written"* (Luke 4: 4, 8, 10); three times He appealed to the Word of God and His victory over evil was assured. He was strong and the Word abided in Him, and as a result, He overcame the wicked one:

This is a thrilling assurance. I think that those who are reading John's Epistle couldn't help, at this point, but be encouraged because to be reminded that our sins have been forgiven, that we have known the one who is from the beginning, and we have known the Father, and that we are strong, and the word of God abides in us, and we have overcome the wicked one. Why, that's marvelous, and at the heart of it is the Lamb, the Word of God, the Word of God abiding in them.[2]

• Ye have Overcome
I have written unto you, young men, because . . . ye have overcome the wicked one (1 John 2:14).

Satan has no power over you! Christ defeated Satan at the cross. We Christians, through the Spirit, defeated Satan the moment of our regeneration, made spiritually strong out of natural weakness. We achieved the first grand conquest over him when passing from darkness to light, but afterwards we now need to maintain a continual keeping of ourselves from his assaults. But be aware! Satan is still alive and well, and active in our nDNA, *as a roaring lion, walketh about, seeking whom he may devour* (1 Peter 5:8).

These are youths in the prime of their spiritual lives, valiant soldiers fighting under the banner of Christ, who had confounded Satan in his wiles and overcame him by the blood of the Lamb; confirmed disciples of Christ; persons who were well-grounded in the truth and thoroughly exercised in Christian warfare. They were persons in the prime of life and in the zenith of their faith and love; over-comers in many battles by the power of faith.

In the eyes of God, the "young men" in Christ are seminally and representatively:

• In His Loins
And as I may so say, Levi also, who receiveth tithes, paid tithes in Abraham. For he was yet in the loins of his father, when Melchisedec met him (Heb. 7:9–10).

We have discussed these verses in some detail previously where we have undisputedly claimed that every Christian believer is both in the loins of Adam as pertaining to sin, and also in the loins of Christ as pertaining to salvation:

> Just as Levi was "*in* Abraham," not only seminally but representatively, so every one of God's children was "*in Christ*" when He wrought out that glorious work which has honored and pleased God high above everything else. When the death-sentence of the law fell upon Christ, it fell upon the believer, so that he can unhesitatingly say, "I was crucified with Christ." So too when Christ arose in triumph from the tomb, all His people shared His victory. When He ascended on high, they ascended too. Let all Christian readers pray earnestly that God may be pleased to reveal to them the meaning, blessedness, and fullness of the words "In Christ."[3]

And since we are in His loins, able to assume responsibility for actions well before our birth, events that have occurred in the past, over 2,000 years ago at the cross of Calvary, we as Christians can lay claim to the reality that we have been:

• Baptized into, and have Put On, Christ

For as many of you as have been baptized into Christ have put on Christ (Gal. 3:27).

This is not water baptism being spoken of here, a public profession that we have received Christ as our Lord and Savior, but spiritual baptism by the Holy Spirit into one body, the Body of Christ. By baptism, we all enter into union with Christ, united with Him at regeneration and now *have put on Christ*— have received his Spirit, entered into his interests, and copied his manners as if to be *clothed* with Christ:

> To put on, or to be clothed with one, is to assume the person and character of that one; and they who do so are bound to act his part, and to sustain the character which they have assumed. The profession of Christianity is an assumption of the character of Christ; he has left us an example that we should follow his steps, and we should, as Christians, have that mind in us which was in him.[4]

To be clothed with a person is to *put on* the interests and relationships of that person, to enter into his views and thus to imitate him, and to be wholly in agreement with him. In Greek phraseology, to *put on* means "such a one hath put on such a one," that is, they clothed themselves with him, sinking into his own garments:

> Let us put on decent garments – let us make a different profession, unite with other company, and maintain that profession by a suitable conduct. Putting on, or being clothed with Jesus Christ, signifies receiving and believing the Gospel; and consequently taking its maxims for the government of life, having the mind that was in Christ.... the phrase putting on the shechinah, or Divine majesty, to signify the soul's being with immortality, and rendered fit for glory.[5]

We *have been baptized into Christ* is a past event, evidenced by the past tense of the Greek verb. The action of the verb is an accomplished event, a joyful event, and a Christ-honoring event; therefore, *put on Christ* in its entirety; in possession of immortality and fit for glory. Be admonished to live up to that honor.

• **Crucified with Christ**
I am crucified with Christ: nevertheless I live; yet not I,
but Christ liveth in me (Gal. 2:20).

The old man in us is continually being crucified but the
new man, though we live in the flesh, yet we do not live after
the things of the flesh, for we live by faith in Him who died,
and gave Himself for us. Even in this mortal body, we live by
faith in the Son of God. Christ is our treasure, and where our
treasure is, there is our heart; for to believe in Christ crucified
is to believe that we were crucified with Him.

• **Ye are Dead in Christ**
For ye are dead, and your life is hid with Christ in God
(Col. 3:3).

How does it feel to you to be dead in Christ? I love it! I may
still be alive in this world, but the fact of the death of Christ is
proof positive that there is no hope of salvation for anyone other than by what He accomplished at the cross at Calvary. Life
is only possible through Him, for it is not I that is living at this
moment, but I live because of the life of the One who lives in
me. Praise the Lord, my *life is hid with Christ in God!*

• **Ye are Risen with Christ**
If ye then be risen with Christ, seek those things which
are above, where Christ sitteth on the right hand of God.
Set your affection on things above, not on things on the
earth (Col. 3:1–2).

Do what Paul suggests; *Set your affection on things above,*
not on things on the earth. Why go around doing the things
that other people do, things that are not counted for righteousness. Remember, *put on* Christ, that others may wonder
what on earth is the matter with you.

• **Ye are Seated in Heaven**
And hath raised us up together, and made us sit togeth-
er in heavenly places in Christ Jesus (Eph. 2:6).

This is our heavenly position and it has already occurred!
We as humans in time do not possess the divine attribute of

omnipresence but, nonetheless, we are already seated in heaven with Christ. As Christ was both in heaven and on the earth at the same time, so we also should enjoy and proclaim this position, which is not imaginary but actual.

CHAPTER THOUGHTS

The Christian at the embryonic phase has now grown up into a magnificent child of God, accumulating all of the previous marks ascribed to him, and is now ready to assume the status of *fetus*. He has passed through the birth phases as a zygote, a blastocyst, and now as embryo, retaining the divine marks that identify him as such: first, as newborn babes in Christ—the elect, with sins forgiven, and possessor of incorruptible seed; second, as little children—ye know the Father in an intimate relationship, as a chimera of two natures, a priest in the temple of God, and as our friend; third, as embryos—strong, as overcomers by the Word, baptized, crucified, dead, risen, and seated in heaven.

Justification pronounces the believer free of the guilt of sin. Christ paid it all. The sin pathogen is not to be found in our sDNA flesh, but due to the chimeric nature of our being, nDNA flesh is constantly in need of forgiveness. Every molecular cell in our body, not as yet regenerated by the Holy Spirit, cries out for the election of the whole body, yearning for the image of God to be made manifest in each cell that it may be of incorruptible seed because:

- Of our position as a royal priesthood established by heredity.
- We were in His loins prior to birth, awaiting our conception.
- We have the seed of God tucked away in our cellular structures.
- We display those characteristic traits of Love, Joy and Peace in the cell, gifted by the Holy Spirit, expressed by *the red color trait in the bloom of a flower,* or as *the light or dark-winged traits of a butterfly....*

for indeed, what are "Y-chromosomes," and what are "loins," and what is "seed," and what is "fruit," but His sDNA, the Body of Christ. Is it any wonder then that the Bible comes alive as never before as we contemplate the amazing relationships between Christians and the man, Christ Jesus? For if Christ be crucified, dead, buried, and raised from the dead, then we also partake in these events both here and *now*, as His sDNA is our sDNA, and *where He is, we are!*

All that is asked of us to do is to believe in Him, obey his commandments, and to love all the brethren. And we didn't have to do anything. We didn't have to go all the way to the cross and hang there all day and suffer a humiliating death. We are crucified and dead because He was crucified and died; we are raised and seated in heaven because He was raised and seated in heaven. How easy is that for us to do, to simply ask for forgiveness and receive Him as our Friend. What more can a Christian ask for?

TIDBITS OF PART V
- God is Light:
  - o Jesus Christ;
    - ◆ The Informer.
    - ◆ The Light of the World.
  - o The Light that shines;
    - ◆ In the darkness.
- Light is photons:
  - o Carrier of Information;
    - ◆ Knowledge.
- Conceptual Thought:
  - o Messages from Jesus.
- Man:
  - o Fearfully and wonderfully made.
- Let me count the ways:
  - o Genome;
    - ◆ 3×10^{-9} nucleotide base pair letters.
  - o Length;
    - ◆ 1.02 m long.
  - o Volume;
    - ◆ 3.19×10^{-9} mm^3.
- Genomes fitting on the head of a pin:
  - o 1.31×10^9.
  - o Need 3.57×10^9 equivalent CDs;
    - ◆ To equal capacity of all genomes on pinhead.
- The Emotional God:
  - o The fruit of the Spirit;

- ◆ Love, Joy, Peace, Longsuffering, Gentleness, Goodness, Faith, Meekness, Temperance.
- FRUIT→sNeuropeptides→sEmotions→sImune Cells:
  - o LONGSUFFERING →s GABA → sLongsuffering → sT Cell.
  - o GENTLENESS→sDopamine→sGentleness→sB Cell.
  - o GOODNESS→sEpinephrine→sGoodness→sNK Cell.
- Longsuffering, Gentleness, Goodness:
  - o Verses – Uncleanness, Seditions, and Witchcraft.
- Phase 3→*Justification* →*Organismal* →*Embryo* →*Young Men:*
  - o Justification;
    - ◆ Just as if I never sinned.
  - o Organismal;
    - ◆ Like the red color of a flower's bloom.
  - o Embryo;
    - ◆ Christian character emerges.
  - o Young Men;
    - ◆ Strong.
    - ◆ Overcomers.
    - ◆ In His loins.
    - ◆ Ye have the Word.
    - ◆ Baptized, put on, and dead—in Christ.
    - ◆ Crucified, risen—with Christ.
    - ◆ Seated in Heaven—with Christ.
    - ◆ Created in the image of God.

PART VI

The Battle for the Spirit

19

LIFE'S AMAZING STORY

"God is eternal life" (Romans 6:23)

If the source of *Empowering Energy* is identified as God the Father, to supply life-transforming power from an infinite reservoir of energy to an nDNA molecule, to transform into a spiritual sDNA molecule; and the *Conceptual Information* is identified and resident in the mind of Jesus Christ, the Word of God, to embed into the nucleotide sequence of the sDNA molecule the code containing specific instructions for leading a spirit-filled life; then the *Converter* mechanism, the third requirement for suspension of the Second Law of Thermodynamics in order for a divine miracle to occur, is the miraculous process of *Regeneration*—the bringing together and the binding of the enormous energy and power expended by the Father, with the infinite informational content latent in the Son—a work exclusively and completely performed by God the Holy Spirit, the Omnipresent, for the Almighty Spirit quickens, indwells, and sustains every believer with His eternal Life:

> *The Spirit of God hath made me, and the breath of the Almighty hath given me life* (Job 33:4).

THE LOVE OF GOD: THE FATHER

There are three major works that are absolutely necessary to be performed by the members of the Trinity, prior to the initiation of the joy of the new birth, spiritual work to be done in the overall plan in the salvation of a sinner.

The first absolute necessary act in the initial work performed by God the Father, the Holy Mutagen, is in the *inward* love of God in the preparation of the body to affect a new act of creation, by attacking the nucleotide structures of the predestination genes that start the process of election. The nDNA structure of a sinner is riddled with sin and thus aligned improperly. The helix is unzipped and rearranged to conform to the life of each individual person; mRNA produced; and then zipped back up again. The *outward* love of the Father is in the visible manifestation of the new birth, of the inward agape Love, a new Love permeating the soul of the elect.

THE LIGHT OF GOD: THE SON

The second absolute necessary act in the overall plan of salvation is the work of *sacrifice* that is characterized by the phenomena of Light. There are two manifestations of light; inward light and outward light. Both are as different as the East is from the West; as deep and wide as the ocean is from the dew and the raindrop.

The light that lights our path is the *outward* light that guides every soul to Christ. This work by God the Son is accomplished in the death of His body on the Cross at Calvary. It is the path taken in the redeemed life of the believer.

The light that shines in our hearts is the *inward* light that permeates the entire body, infusing information, knowledge, and understanding to the soul. He it is who is the master Architect, the One who has within Himself the plans of life, and how they are to be embedded within the sDNA sequence; of how to select, who selects, and which genes are selected; which nucleotide base pairs are required—*exons*, and which nucleotide base pairs are not—*introns*; what proteins are to be made for each individual person to express special transcription and translation operations; to form unique amino acid configurations, and to produce unimaginable proteins never before existing in the body; and how to *fold* each protein so that millions will fit on the head of a pin. Information latent in the Son is the real secret of life that expresses itself in outward appearances.

THE LIFE OF GOD: THE HOLY SPIRIT

The third absolute necessary act in the overall plan of salvation is the work of *regeneration*. This initial work by God the

Holy Spirit is an act of revitalization of a dead body. nDNA is a dead molecule for we are dead in our sin, but: *And you hath he quickened, who were dead in trespasses and sins* (Ephesians 2:1). A dead molecule cannot resurrect itself to a new life, to a new creation. It requires power and information resident.

THE INWARD LIFE

It all goes back to the Garden of Eden. Every time we endeavor to speak of the body, the soul, or the spirit, we inevitably find ourselves back in the book of beginnings; in the book of ultimate importance and purpose, the book of Genesis where it all began and where it all will end. The Holy Spirit used the conversion mechanism of *generation* to give life and to vitalize the very first nDNA molecule in Adam. Since the Second Law of Thermodynamics took effect after the fall of Adam, that life suffered under the penalty and curse of sin. But in His infinite love and wisdom, God has provided a means for a New Life, a *re-generation* of the original life; a new birth; a spiritual birth similar to physical birth; where we play no active roll. Re-birth is a new creation, the calling into existence of something different, something entirely new from what was there before. Here, Jesus instructs Nicodemus:

> *Verily, verily, I say unto thee, Except a man be born again, he cannot see the kingdom of God. Nicodemus saith unto him, How can a man be born when he is old? can he enter the second time into his mother's womb, and be born. . . . Marvel not that I said unto thee, Ye must be born again* (John 3:3–4, 7).

Can a person grow up in a Christian home where both parents profess to be Christians, in a home where Jesus is acknowledged as Savior, go to church regularly on Sunday mornings and to Bible study on Wednesday evenings, hear the Good News message of the Bible proclaimed, and still not be Christian? Is this possible? Absolutely it is! But you may inquire—"Why can't good deeds and obeying the Law of Moses grant you admission to heaven?" The answer to this question is so simple. Listen to Jesus—*"ye must be born again."*

Regeneration is the sovereign work of God the Holy Spirit. The initiative is with Him, not with us. We are incapable of performing an act of re-birthing ourselves because of the fact that regeneration *precedes* faith! If faith came *before* regeneration, man would be in a position to save himself, and Scripture would abound with acclamations of grandeur for a work well done in the flesh. In this scenario, man, out of the goodness and tender mercies of his own heart, usurps the work of God and raises himself up into new and glorious spiritual heights. He is now in full control of his destiny. All that is needed is a little spot of self-righteousness, a little bit of divine power, a little morsel of deity in his soul. God is not needed, not required, and therefore, not a factor.

But is this the actual case? Does man possess the power to rearrange the nucleotide sequences in his predestination genes? Does he have the knowledge to embed within the new arrangement a new concept of codes that, when activated, will lead to a new spiritual life within his fleshy body? Can he perform this miracle without the assistance of God? Is God incapable of achieving the desired results of a new birth? If so, can any Arminian explain to all of us how he did this without God?

It is by the power of the Holy Spirit, and *only* by the power of the Holy Spirit, by no other power under heaven, whether by man, by angel, or by Satan himself, can ever affect a salvation as described in Scripture. It is He who is the Author, the Sustainer, and the Finisher of our faith. Regeneration is monergistic in its initial stage but progresses synergistically in its final stages. We are spiritually dead in the initial stage and thus cannot cooperate; but spiritually alive in the final stage, as seen in the *accumulation* of faith and of good works towards our nDNA flesh.

• The Necessity of Regeneration

Man has forfeited his original created life. No longer is he able to walk in the cool of the Garden and commune with God. He has lost his innocence and now is under the penalty of death. What can man do to regain that privilege, to once again communicate with his God, and to live a normal life as He intended him to live? The effects of sin are enormous; his life is of no affect; what is needed is a new life, a new birth.

There is no other choice, no other remedy, no other solution: *you must!* Nicademus at first didn't understand what Jesus was talking about when he asked: *How can a man be born when he is old? can he enter the second time into his mother's womb, and be born?* (v. 4). What was it about "must" that Nicodemus did not understand? In his own mind, he was contemplating a solution to the problem of sin by the utilization of his own power, perhaps by working synergistically with God. But in God's economy, this is an impossible task.

Generation implies natural birth, regeneration implies spiritual birth; but it is evident that man cannot distinguish between biological birth and spiritual birth, between flesh and spirit. Self regeneration of the soul is impossible to perform in its own power. All ambiguity evaporates with the word *must*. Man in his fallen state is totally depraved, unable to satisfy a righteous and holy God. It is possible for him to be mature, intelligent, ambitious, and successful, but he is incapable of exercising the proper thoughts relating to his destiny, and is void of all concepts of divine intervention. His mind is set against God and every notion of his will is corrupt and sinful. There are no exceptions; flesh begets flesh; utterly impotent, which means that he cannot beget that which is spiritual.

Regeneration is a mystery, a secret act of God in time; once being dead, now we are alive. It is a matter of life, not of the life that we were born with, but to possess the life of God, which requires another birth.

It is clearly seen that man has no participation in the regeneration of the Adamic life, for regeneration requires, not only personal forgiveness of sins, but also of a *death*, not our death, but the death of His Son, Jesus.

You may, if you desire, if you think that you possess the power to affect your own regeneration by forgiving your own sins, then try hanging on a cross until you succumb of suffocation. But I warn you, this is not an easy task because you have to die before regeneration takes effect. But, if you are dead, then you are dead! (I strongly suggest that you don't try this).

• The Power of Regeneration

The story of the caterpillar reminds me of my previous condition, when I enjoyed eating the bitter green leaves of sin and loving every moment of it. All Christians can confess to this. Consider the following, determine in your own mind if butterflies are still basically caterpillars, that is, if butterflies retain and possess the same qualities and life practices as that of the caterpillar:

Many butterflies migrate over long distances. Particularly
famous migrations being those of the Monarch butterfly
from Mexico to North America, a distance of about 4,000
to 4,800 kilometers (2500-3000 miles). Many butterflies,
such as the Monarch butterfly, are migratory and capable
of long distance flights. They migrate during the day and
use the sun to orient themselves. They also perceive po-
larized light and use it for orientation when the sun is
hidden. . . . Like many other members of the insect world,
the lift generated by butterflies is more than what can be
accounted for by steady-state, non-transitory aerodynam-
ics. Studies using *Vanessa atalanta* in a windtunnel show
that they use a wide variety of aerodynamic mechanisms
to generate force. These include wake capture, vortices at
the wing edge, rotational mechanisms and Weis-Fogh
'clap-and-fling' mechanisms. The butterflies were also
able to change from one mode to another rapidly.[1]

Can the caterpillar do all that? Can he lift up all his legs
and effortlessly soar through the air as a butterfly does so majestically with his wings? You say no, and rightfully so. Then
the larval caterpillar is no longer a caterpillar; it is now a butterfly by the action of metamorphosis, a new creation, a new
creature, a new species! And so are you if you believe on the
power of the Holy Spirit to affect a salvation in your heart.

The power expended by the Holy Spirit in the life of a
Christian is enormous. Once I was a sinner, an infidel, wallow-
ing in my own vomit; and now, by the grace of God, I am
raised up unto heights unknown; soaring as a new-born butterfly, where even angels fear to tread. I was so happy in my
previous condition, loving sin, enjoying every moment immensely, associating with those who practiced the same, they
also rejoicing in it. I was dead in sin, didn't care if I sinned,
couldn't wait until I sinned again, and didn't know that I was
sinning, but continually kept on sinning.

There was nothing good in me. I didn't know that at the
time. I thought that I was good, kind, and benevolent. I ate
from the bitter green leaves of sin, loved eating it, and could
not get enough of it. Little did I know or understand that just
under my skin there were tremendous things going on, things
that would baffle the mind if I just could see it. First, there
was a re-arranging of my predestination genes. I didn't know
that I had predestination genes. Do you know that you have

predestination genes? With all the scientific study going on today, someone should have told me that I had predestination genes. Maybe I could have done something spectacular with them. Where did these genes come from? Who turned them on? Why were they turned on? What does it mean that they were turned on? I confess that I don't know much about all this and why it is all so necessary for me, but one thing I do know, that it has something to do with *ye must be born again* (v. 7).

For whom he did foreknow, he also did predestinate to be conformed to the image of his Son (Rom. 8:29).

When you look in the mirror, what do you see? If you are like me, and you are like me, you will see a body, a single body, a single Christian body. But just below the thin surface layer of the skin, there are two entities, one of flesh from our *fleshly* parents, natural, *that which is born of the flesh is flesh,* and one of flesh from our *spirit* parents, spiritual, *that which is born of the Spirit is spirit* (John 3:6). This verse is a sure proclamation that everyone born of fleshy parents is of nDNA flesh, and everyone born of the Spirit is of sDNA flesh.

My thoughts go back to the plight of the lowly caterpillar, eating bitter green leaves, while loathing the sweet aroma and delicious nectar in the bloom of a flower. He was perfectly satisfied with the condition of his life, a voracious insect capable of destroying fields of healthy and thriving plants, willing not to change his way of life, his "modus operandi."

The caterpillar didn't know it, had never entered into his mind that all that was about to come to a climatic end. He didn't know that he was now becoming sleepy, sluggish, and full of undigested green leaves. He looked for a place where he could curl up, yawn a few times, and take a nap for a while. He clung to the bottom of the same leaves that he was eating and went to sleep; it was very refreshing, he thought, but never knew what was about to happen to him while he was sleeping.

But when the lowly caterpillar woke up, he was again very hungry. He tried to eat some of those delicious green leaves that he munched on all his life but he quickly found out, to his dismay, that he had no teeth to bite off a chunk. "What happened to all my teeth?" he murmured. Then he decided to go for a walk to try to find his missing teeth, but instead of moving all his feet he suddenly found himself moving his wings—

he was flying! "Where did my feet go?" he murmured and "what is this flying that I'm doing?" And wouldn't you know it, this thing called an "antenna" guided his path and he landed on the pedal of a beautiful flower. Suddenly his "proboscis" started coming out of his head ("his what?" he thought), and then, protruding from his head it started sucking up the sweet nectar of the bloom. And was it ever so sweet!

Now he knows what happened to him. The lowly caterpillar experienced a spectacular metamorphosis, a divine change from one form of flesh into another form of flesh, from caterpillar flesh that crawls on the ground, a voracious eating machine that mutilates and destroys the greenery, into butterfly flesh that flies through the air sporting colorful wings, a beautiful, delicate, and sweet nectar-sucking flesh—a new creature, and it happens all the time, all day, every day.

THE OUTWARD LIFE

The outward workings of the Holy Spirit, producing visible marks of the newly regenerated Christian, are manifold:

• Called

Moreover whom he did predestinate, them he also called: and whom he called, them he also justified: and whom he justified, them he also glorified (Rom. 8:30).

There is an outward call by the Holy Spirit and there is an inward call. Everyone who has ears to hear hears the outward call. It is a calling designed to bring everyone to faith in Jesus Christ, a calling that, if acted upon by the hearers, will usher in the process of regeneration of the soul. Both the non-elect and the elect hear the outward call loud and clear. It is evident, however, that the non-elect will not respond to the outward call and have made it of no affect, rendering the call ineffectual. The inward call is heard by the elect and act positively upon it, rendering the call effectual. Everyone who hears the ineffectual outward call does not respond, will not respond, and cannot respond. They cannot hear inwardly!

Gleaning from the brief discussion of the effectual calling by the Holy Spirit, we have previously stated that everyone has predestination genes, but not all predestination genes are the same. Some predestination genes are common or generic, that is, *ineffectual* in salvation. These are the majority of genes that determine all aspects of life in the flesh. They are the genes that we have been calling natural genes, having been turned on at conception and are the genes that are studied today in the laboratories around the world. They are common predestination genes that conform each of us to who we are in particular and to no one else. They perform well in the natural realm.

In contrast to ineffectual predestination genes are the *effectual* predestination genes that, when called to be turned on, will never be turned off, and will ultimately and surely conform each of us to the image of His Son, and to no one else.

• Drawn
No man can come to me, except the Father which hath sent me draw him: and I will raise him up at the last day (John 6:44).

How can anyone explain the drawing of the Spirit better than that of Adam Clarke:

> But how is a man drawn? . . . A man is attracted by that which he delights in. Show green herbage to a sheep, he is drawn by it: show nuts to a child, and he is drawn by them. They run wherever the person runs who shows these things: they run after him, but they are not forced to follow; they run, through the desire they feel to get the things they delight in. So God draws man: he shows him his wants-he shows the saviour whom he has provided for him: the man feels himself a lost sinner; and, through the desire which he finds to escape hell, and get to heaven, he comes unto Christ, that he may be justified by his blood. Unless God thus draw, no man will ever come to Christ; because none could, without this drawing, ever feel the need of a saviour.[2]

If you are a Christian, then you have been drawn!

• Enlighten
A light to lighten the Gentiles, and the glory of thy people Israel (Luke 2:32).

You can read the Bible day and night to gain some insight into spiritual things, but the acquired knowledge is but a natural knowledge. You can pray twenty-four hours a day, go to seminary and study until the cows come home, preach from the largest pulpits in the country, but you still will not know anything about spiritual things. It is the Holy Spirit, and Him only, who supernaturally enlightens the mind:

> The same law which controls a planet affects a grain of dust. As God caused revelation to arise gradually, and, growing clearer and clearer, to become clearest when it seemed about to expire, so in the experience of each individual, the dawn preceded the day. When the light of divine grace first visits a man, it shines with feeble beam. Man by nature is, like a house shut up, the windows of which are all boarded over. Grace does not open every window jet once and bid the sun stream in upon weak eyes accustomed to darkness. It rather takes down a part of a shutter at a time, removes some obstruction, and so lets in, through chinks, a little light, that one may be able to bear it by degrees. The window of man's soul is so thickly encrusted with dirt, so thoroughly begrimed, that no light at all can penetrate it, till one layer is taken off, and a little yellow light is seen; and then another is removed, and then another, still admitting more light, and clearer. Was it not so with you who are now walking in the light of God's countenance. Did not your light come to you by little and little?[3]

• Indwelt

A new heart also will I give you, and a new spirit will I put within you: and I will take away the stony heart out of your flesh, and I will give you an heart of flesh. And I will put my spirit within you, and cause you to walk in my statutes (Ezek. 36:26–27).

The possession of the Holy Spirit is the distinguishing mark of a Christian, for to be without the Spirit is proof positive that we are none of His, for the Holy Spirit takes residence within the body of a Christian, more specifically, within the body that contains spiritual flesh. A new décor is put in place and the new body assumes a new vitality. A *new heart* is created by the Father and a *new spirit* is formed by the Light of the Son that shines in the heart of the new creature. The stony heart of flesh is molded into a new heart, one attuned to the Spirit. Adam Clarke writes of the interchange between the new heart and the new spirit:

> I will change the whole of your infected nature; and give you new appetites, new passions; or, at least, the old ones purified and refined. The heart is generally understood to mean all the affections and passions. I will renew your minds, also enlighten your understanding, correct your judgment, and refine your will, so that you shall have a new spirit to actuate your new heart. That heart that is hard, impenetrable, and cold; the affections and passions that are unyielding, frozen to good, unaffected by heavenly things; that are slow to credit the words of God. I will entirely remove this heart: it is the opposite to that which I have promised you . . . one that can feel, and that can enjoy; that can feel love to God and to all men, and be a proper habitation for the living God. . . . I will put my Spirit, the great principle of light, life, and love, within you, to actuate the new spirit, and to influence the new affections and passions; that the animal spirit may not become brutish, that the mental powers become not foolish.[4]

• Sealed
In whom ye also trusted . . . . in whom also after that ye believed, ye were sealed with that holy Spirit of promise (Eph. 1:13).

Sealing signifies security of ownership of the believer who has placed faith in Jesus Christ, a finished transaction. He is sealed in the position of never being able to lose that coveted position of "Christian," regardless of what may transpire and by whom the attack may come. The believer is sealed by, in, and of, the Holy Spirit and is thus eternally secure.

CHAPTER THOUGHTS
Each of the members of the Trinity has a job to do: the Father—to expend enormous amounts of energy to turn-on the predestination genes lying dormant in the genome; the Son—to furnish the vast amount of information, the intelligence necessary to energize the newly formed genome; and the Holy Spirit—to bring these two endeavors to fruition through the act of regeneration, to function as a spiritual Christian—one

consciousness with two bodies—one with sDNA and one with nDNA flesh.

Each of the members has an inward and an outward endeavor in the performance of their tasks: the Father—to spread forth the inward love of God in the preparation of the genome for election; and the outward love, the visible manifestation of the mercy and love of God; the Son—the inward light, the light that shines in our hearts; is the light that permeates the soul, infusing intelligence and knowledge to non-living molecules to act intelligently in gene expression; of translation and transcription of the code inherent on the mRNA molecule; and the outward light, the light that shines in the darkness; the light that lights our path, that light that guides us to consider the benefits of the Cross; and the Holy Spirit—administers the inward life of regeneration; the implantation of the "new life," the "new man," the "new creation," a sovereign act of God; and the outward workings produce visible marks of the newly regenerated Christian—the calling, the drawing, the enlightening, the indwelling, and the sealing.

FAITH, MEEKNESS & TEMPERANCE

*"But the fruit of the Spirit is ... faith, meekness, temperance;
against such there is no law"* (Galatians 5:22–23)

Allow me to divide life into three main groups: thoughts, emotions, and feelings. Thoughts stimulate the production of neuropeptides, and in turn, neuropeptides produce emotions, and emotions produce feelings. Put in another way; we feel what we think by means of our emotions.

We see in Scripture a changing progression of spiritual and biological activity. In Genesis 2:7, in the original creation, we read that the lineup is, body–spirit–soul, whereas in the N.T. we read, in 1 Thessalonians 5:23, spirit–soul–body, in the ongoing life of a Christian. We now offer another arrangement, namely, Holy Spirit–spirit–body–soul, where the Holy Spirit dwells only in the spirit of man, communicating with him in the spirit; the spirit, in turn, receiving and processing His messages, passes them along to the appropriate receptors in the body via neuropeptides; which, in turn, produce emotions in the body and feelings in the soul. The body does not feel emotions; the body produces emotions; emotions then generate feelings; and the soul feels all the feelings. There is some sort of biological connection between emotions and feelings, and between body and soul.

Emotions and feelings do not occur unless neuropeptides bind to the appropriate receptors that ultimately call emotions into the subconscious mind, and then feelings into the conscious mind. Positive or negative thoughts will produce neuropeptides that will react in a positive or negative manner respectively—in a positive manner to produce positive emotions and feelings, and in a negative manner to produce negative emotions and feelings; all depending upon the abundance of binding peptides and receptors, or the lack of the same.

A thought creates a neuropeptide, a neuropeptide elicits an emotion, and an emotion produces a feeling—that is why we feel the way we feel. If we do not like the way we feel, we may simply start to think of pleasant things in our lives, of feelings that we like to feel, and keep thinking about those good feelings. In time, we will rewire our own brains *on the run* with new neuronal pathways that will maintain those pleasant emotions and feelings—as long as you don't start thinking about bad ones again. This will produce positive and negative emotions that balance out, producing a resultant feeling:

> Why would emotions be balanced, why not just have only positive emotions? Because if you are curious, your curiosity is going to backfire when there is a failure (you'd be curious in a failure). Or if you are overly surprised, you would be just as surprised at a bad thing happening as you would as a good thing happening, leading to being happy and sad. Or if you got angry at something, you are then likely to become pleased by the opposite thing happening, so the emotions tend to balance out.[1]

Not so in the spiritual world; there are only positive emotions dispensed by the fruit of the Spirit. On the other hand, within the realm of all positive spiritual emotions, there is a balancing act between the various spiritual emotions that will produce a desired spiritual effect, a final level specific for each of us that will ultimately distinguish each of us as a unique spiritual person.

GOD'S THOUGHTS

Thought is *information* and information drives everything in the human body. There is information in all energy and matter in the universe. Radiation from the sun contains information concerning heat, light, color, velocity, etc. We have receptor receivers in our body that will detect some of these quantities. Information does not emanate from the receptors but from the matter itself. The receptors receive and process the information, sends it forward into the cell, into the nervous system, and ultimately to the brain and beyond.

We are primarily informational beings, for without thought processes we cannot function. Specific thoughts generate specific neuropeptides; these in turn seek specific receptors to communicate with, and to advance the informational content of the thought to other areas of the body; until the entire thought is manifest and actions generated. Specific receptors located in every corner of the body will read and measure information associated with each thought impressed upon message-carrying neuropeptides.

Thoughts and emotions "talk" to each other in their own language and to the billions of immune cells in the body. We can now begin to recognize the inter-connectedness of our thoughts, emotions, and immune cells. Our thoughts are not simply philosophically connected to our emotions, but are actually biochemically connected. Neuropeptides are not merely an artifact of thought, or emotion, or feeling; they *are* the thought, or emotion, or feeling.

The Holy Spirit communes with man by His thoughts, ultimately entering the cerebral cortex of the brain, the seat of cognition where thought, memory, and consciousness are processed and interpreted. His thoughts enter the gray neuronal matter by selecting specific receptors located on the surfaces of dendrites; producing neuropeptides that transverse the synaptic cleft, and begin their journey toward the cell nucleus— where they are received, processed, and retransmitted along the white axonal matter to various locations in the body:

> Every second of life, a massive informational exchange is occurring in the body, between the neural, the hormonal, the gastrointestinal, and the immune systems, each with a different tone, humming a signature tune, rising and falling, waxing and waning, binding and unbinding, and if we could hear this music then the sum of these sounds would be the music that we call emotions.[2]

Devine messages do not enter into the realm of human material, nor do they become part of the human body. His thoughts are still His thoughts—immaterial, informational messages. They do not penetrate the cell nor will they. The divine will not become part of human flesh. The thought docks to the receptor but never enters further, producing a directional informational exchange into the cell.

> Thanks to messenger molecules (neuropeptides), events that seem totally unconnected- such as a thought and a bodily reaction- are now seen to be consistent. The neuropeptide isn't a thought but it moves with thought, serving as a point of transformation. . . . A neuropeptide springs into existence at the touch of a thought, but where does it spring from? . . . A thought of fear and the neuro-chemical that it turns into are somehow connected in a hidden process, a transformation of non-matter into matter.[3]

We are spiritual beings in a physical body with the erroneous perception that we are separated. The reality is that we are organically and biologically connected by means of peptides and receptors, producing emotions that oscillate back and forth, connecting the physical with the spiritual within us.

The prophets and the apostles wrote the Scriptures with their own vocabulary of words, obtained throughout a lifetime of living, but their thoughts were *moved* (Gk, *phero*, to bring forth), i.e., brought forth by the Holy Spirit:

> *For the prophecy came not in old time by the will of man: but holy men of God spake as they were moved by the Holy Ghost* (2 Pet. 1:21).

We are commanded to witness to our unbelieving friends of the beauty, of the glory, and of the benefits of God our Savior. We do so at the populational level where we talk to one another face to face, a one-on-one conversation. But what of the witnessing that takes place at the molecular level between the "new man in the new nature" with the "old man in the old nature?"

Cells are in constant communication with each other. Spiritual sCells are constantly talking with natural nCells, the good cells witnessing to the bad cells. Thoughts of love, joy, and peace flow freely in one direction, crossing the boundary from the spiritual to the natural, the boundary that marks-out the new man from the old man. Thoughts communicated from the new man are the thoughts obtained from the Holy Spirit; thoughts falling on the enticed mind of the old man seeking regeneration; thoughts that enable him to see the magnitude of his sin and to taste of the Savior; to seek-out the life of the new man in Christ.

SPIRITUAL FRUIT: GROUP 3

Thoughts from the Holy Spirit contained in the fruit of the Spirit ignite in us a spiritual neuropeptide materialism that carries forward the message of God intended for the believer upon regeneration; a message delineating precise instructions as to the progression of one's sanctification; of how neuropeptides and white blood cells are to act and interact; producing in the human body an image of the divine—marks of a Christian that are discernible to all.

The last of the three fruit of the Spirit as it relates to our responsibility toward ourselves are:

• FAITH → sVasopressin → sFaith → sMonocyte
Thy mercy, O LORD, is in the heavens; and thy faithfulness reacheth unto the clouds (Psa. 36:5).

THE FAITH OF GOD

We are talking here about the Faith (Gr. *pistis*) or faithfulness of God. Sound the clarions loud and clear—God is faithful!—for His faithfulness endures forever.

It is of the LORD'S mercies that we are not consumed, because his compassions fail not. They are new every morning: great is thy faithfulness (Lam. 3:22–23).

The refrain lyrics by Thomas O. Chisholm of that great old-time favorite hymn (inspired by the above passage by the prophet Jeremiah in the Book of Lamentations): *Great Is Thy Faithfulness*,[4] speaks beautifully of the Faithfulness that flows only from the throne of God, continually filling the heart with mercies of love, joy, and peace:

> *Great is Thy faithfulness!*
> *Great is Thy faithfulness!*
> *Morning by morning new mercies I see.*
> *All I have needed Thy hand hath provided;*
> *Great is Thy faithfulness,*
> *Lord, unto me!*

• sVasopressin

The biochemistry of falling in love is confined in three stages of Greek thought: *eros* (sexual love), *philia* (euphoric love), and *agape* (eternal love).

Erotic or sexual love is the basic and instinctive stage of falling in love that is driven by our sex hormones, testosterone and estrogen. Philia or euphoric love is the falling deeply and madly in love, an intense emotional attraction driven by two important neuropeptides: serotonin and dopamine.

There are three critical factors for a long and lasting agape love: bonding, trust, and faithfulness. Two neuropeptides seem to be the key in the development of strong bonds between partners: oxytocin and vasopressin. We have seen that oxytocin is the *love* neuropeptide that elicits eternal bonding in the agape Love of God, while vasopressin elicits faithfulness in the bonding pair, hence, *"the faithful peptide."* In order to remain loyal to Christ, we must forever remember who Christ is and what He has accomplished for us, facilitating a role in recognition and memory. As a result, the feeling of bonding, coupled with recognition and memory, leads to eternal faithfulness.

• sFaith

You have read it in print, you have sung it in songs, and you have at times shouted it from the mountaintops, that all the world needs today is love. Love, of course, is fine and noble, and, at times, exhilarating. However, the world is in need of much more; more virtues of wisdom, integrity, and faithfulness. These are to be sought after, difficult to obtain in the natural world, but readily available to the spiritually minded. God possesses all of them in abundance, and one of His greatest attributes is His Faithfulness.

One major theme in all of Scripture, one of God's greatest moral attributes, is faithfulness—extending down to the very ones who are reading this paragraph, for it is stated that: *But the Lord is faithful, who shall stablish you, and keep you from evil* (2 Thessalonians 3:3). The Holy Spirit extends to all the emotional virtue of spiritual faithfulness.

When you are faithful, how do you feel? Is there no feeling at all within you; is there not something gushing up and down within your body that gives you a sense of love for God; joy in pleasing Him; and peace in your heart from the knowledge of

knowing Him? How can natural "works of the flesh"—
nIdolatry—the antithesis of sFaith, the first work in an unbeliever's arsenal directed toward his god, the False Prophet,
withstand such an onslaught?

• sMonocyte

Then we have our forces known as *Phagocytes.* These guys
swallow and digest stuff, sort of like Pac Man, if you're old
enough to remember the video game. They engulf and destroy
foreign invaders. Monocytes are the phagocytes that do their
work in the bloodstream, while macrophages are the phagocytes that stand watch in tissues throughout the body.

Monocytes share the "vacuum cleaner" function of neutrophils, but are much longer lived as they have an additional
role; they present pieces of pathogens to T Cells so that the
pathogens may be recognized over and over again; to insure
that an antibody response be mounted; faithfully attacking,
time after time.

• MEEKNESS → sAnandamide → sMeekness → sMacrophage

*Who is like unto the LORD our God, who dwelleth on
high, Who humbleth himself to behold the things that
are in heaven, and in the earth!* (Psa. 113: 5–6).

THE MEEKNESS OF GOD

Can anyone fathom the Meekness (Gr. *praotes*) or the
depths of His humility when God condescended to give His Son
as a ransom to save those who were by sin condemned to be
eternally separated from His presence; to put-on our frame as
a mere man and hang on a wooden giblet; to suffer an inde-
scribable agony because of His love for those whom He created;
and then to die.

The Meekness of God is His power under control. Laying
aside His restraint for a moment, how quickly could God have
smitten the sinner; how easily could He have consumed us
with the fierceness of His anger. In opposition to this, concerning His glory, how much restraint in meekness is needed for
Him to hold back even a fleeting glimpse of His true glory, in
order that we may, in due time, behold Him as He is.

• sAnandamide

Derived from an ancient Indian-language word meaning "bliss," hence *"the bliss peptide,"* anandamide is a natural cannabinoid, or marijuana-like compound, that acts on certain brain receptors involved in pain and mood. When the body senses pain, anandamide binds to cannabinoid receptors and nullifies the sensation by blocking the ability of the nerves to transmit pain signals.

Anandamide is made in the brain and everyone alive is on a natural anandamide-high every day. Someday soon, for those low in this peptide, who constantly suffers pain, there will be anandamide-based medical treatments for those who can't sleep; who have no appetite; who develop strokes or heart conditions; who suffer asthma or glaucoma; who are too restless or anxious; or who suffer any number of other maladies. There is power in anandamide to nullify these and other maladies; to bring the soul back to its God-given position of strength through weakness, for it is the meek that truly understand and rely on the greatness of the Lord God.

• sMeekness

Seek ye the LORD, all ye meek of the earth, which have wrought his judgment; seek righteousness, seek meekness (Zeph. 2:3).

Jesus Christ is not a wimp, and he does not desire His faithful followers to be wimps either. Spiritual meekness is not an indication of holiness, an act of faithfulness, a badge of position, or an act of humbleness. Meekness comes from truly fearing the LORD God. A meek nature is the believer's realization of weakness and insignificance when compared to the power and glory of the Lord.

Meekness is an emotion that has as its main asset a hungering for righting the wrongs attributed to negative emotions. To be meek is to be strong, strong in the sense of not eliciting harsh and negative responses in retaliation to harmful acts directed against you personally by others; acts that normally would result in outbursts of natural emotions in "the works of the flesh"—*nWrath*—the antithesis of sMeekness, the second work in Satan's arsenal aimed at pleasing the False Prophet. Be in control! Be strong! Do not retaliate! Jesus did not open

His mouth when He was wrongfully accused. He remained strong, determined, and, above all, meek.

• sMacrophage
Macrophage plays a key role in tissue removal of cellular debris in the lungs. Removing dead cell material is important in chronic inflammation as the early stages are dominated by aged neutrophils which are ingested by macrophages. They decide if any passing neutrophil is healthy or has to be ingested. If the neutrophil is healthy, it will fire an appropriate signal to the macrophage. The neutrophil will then remain free, but if the neutrophil is not healthy, it will not be able to signal back to prevent its sure demise.

• TEMPERANCE → sOrexin → sTemperance → sDendritic Cell

The LORD is merciful and gracious, slow to anger, and plenteous in mercy (Psa. 103:8).

THE TEMPERANCE OF GOD
God is "slow to anger" in His Temperance (Gk. *egkrateia*), and "patient" in His Longsuffering, that He may extend to those who are of the called—plenteous mercies and abundant graces—tender mercies unto those whom He loves with sweet graces for evermore:

Will the Lord cast off forever? and will he be favourable no more? Is his mercy clean gone for ever? doth his promise fail for evermore? Hath God forgotten to be gracious? hath he in anger shut up his tender mercies? (Psa. 77:7–9).

NO! NEVER!

• sOrexin
Studies have linked orexin activity from sleep and arousal, to feeding and appetite, as well as to inputs of drugs, smoking, and alcohol—suggesting that orexin neurons may have a role in motivation and reward-seeking behavior; extending into the brain regions associated with reward pathways, exhibiting anti-addiction and craving properties. Addiction is not caused

by the drug, but by biological and psychological activity in response to the drug, leading to numerous cravings.

Orexin is the Spirit's way of providing eternal temperance or self-control against the threat of addiction, obesity, and other disorders associated with dysfunctional reward system processing, hence, *"the patient peptide."* Christians are not to worry in their spiritual cells, as addiction of excess or depravation will never be known.

• sTemperance
He that hath no rule over his own spirit is like a city that is broken down, and without walls (Prov. 25:28).

Spiritual temperance is self-control, control over the whole man, in body and in soul, which enables us to live a victorious life. A person who has self-control is calm, avoids extreme behavior, and exercises self-restraint in actions and speech; a strong, moral restraint which keeps us from every immorality. All Christians should have in themselves a moral power which they are to bring to bear in an instant, thereby controlling emotions; slow to anger, slow to speak, and slow to react. Temperance is restraint, not moderation, absolute prohibition from that which injures.

Temperance is placed last in the listing of the fruit; it is the crowning emotion—all others are under its protection. It controls and keeps down the disturbing forces of the sinful nature that yet remain within us—*nRevellings*—the antithesis of sTemperance, the last in the long list in the "works of the flesh," that honors the False Prophet. Its primary purpose is the preservation of all our virtues.

• sDendritic Cells
Dendritic cells represent the pacemakers of the immune system response. They are antigen-presenting cells that help the immune system decide what to attack or what to leave alone. Their main function is to process antigenic material and present it on its surface to other cells of the immune system.

Dendritic cells are present in tissues that are in contact with the external environment, mainly the skin, the inner lining of the nose, lungs, stomach and intestines. Once activated, they migrate to the lymphoid tissues where they interact with

T Cells and B Cells to initiate and shape the immune response, possessing the ability to stimulate T cell proliferation.

CHAPTER THOUGHTS

Cells communicate with each other: sDNA cells "talk" to nDNA cells. They speak to each other in a strange molecular language. What do they talk about?—in their own language, Christians talk about Christ! nDNA cells are sanctified or set apart—set apart to receive some word from God. In their own way, sDNA cells transfer messages to remaining nDNA cells, to inform the cells that there must be some admission of their sin; knowledge of their separation from a righteous and holy God; and a yearning for their need for forgiveness. Else, why is there in life itself an affinity for the things of God? It is known intuitively that God exists and that man must, in the final analysis, give a full account to Him. If cells do not have specific receptors to bind to, to transmit physical and/or spiritual messages in order to elicit cellular responses, then no one can ever come to Christ.

The Holy Spirit resides in the spirit of man and orchestrates the growth of each Christian, cell by cell, organ by organ. By means of thought processes, He forever embeds in the cerebral cortex of the brain, the fruit of the Spirit—those divine moral attributes of God that express Who He is—that will now function as guide to all things spiritual in the human realm.

He uses emotions as the primary driving force in the human body, for our emotions are pure forms of psycho-energy. This psycho-energy transforms into bio-energy at the physical level through the emotional centers of the brain. This is a quantum event where non-matter, like thoughts or emotions, melts into matter, like neuropeptides and immunes.

The fruit of the Spirit, neuropeptides, and immunes, are inextricably related, for it is said that *Ye shall know them by their fruits,* for the battle that rages within you is between the moral excellencies of the "fruit of the Spirit" and the amoral degeneracy of "the works of the flesh," where *men gather grapes of thorns, or figs of thistles* (Matthew 7:16). Develop the fruit of the Spirit in your life; continue in its perseverance, for it is a powerful force to grow spiritually. Cease, therefore, from gathering thorns and thistles which benefit not.

Keep the neural pathways of your body open; energize your immune system, that neuropeptides may flow unimpeded

throughout the body to seek-out specific receptors to bind to; in
order to maintain the link of communications between cells,
organs, and systems; to form permanent relationships. For all
of this—just look to Jesus; think on Him; speak with Him;
trust in Him; and constantly confess to Him that you love Him.
Only then will you notice a change in your body, in your soul,
and in your spirit, that will enable you to leap with love, joy,
and peace unspeakable.

Either way or by any other method, once the receptors are
occupied, no other cell type can gain admittance until that
peptide gives up its position and frees the receptor to mate
with other peptides. In this wise, no infectious virus, no opioid
neuropeptide, or no unhealthy immune may enter into a regenerated elect cell to cause a loss of salvation! Perseverance
of the Saints, the "P" in *T.U.L.I.P.S.,* is thus forever maintained.

21

THE NEW BIRTH FETUS: *THE FATHERS*

"I write unto you, fathers, because ye have known him that is from the beginning" (1 John 2:13)

No longer is it advantageous to remain an embryo, simply sipping on the pure milk of the Word of God, delighting in the initial prospects of your position in Christ: of being effectually called-out; regenerated to newness of life; a new creature; a child of God; reconciled by His shed blood; justified as if I never sinned; and now sanctified and set apart in true holiness before the entire world to see. It is now time in our spiritual growth process to begin eating and digesting the strong meat of the Word, that we may not remain stunted—a Christian dwarf—but to earnestly continue that we may reach the full stature of Christ in us.

One of the greatest lessons that I have been blessed with as I struggled to write this book throughout these many years is wrapped up in the little known phrase in Galatians 3:27, admonishing Christians to—*put on Christ*—be clothed with Christ; sink into His own garments; emulate Him; receive his Spirit; enter into His interests; copy His manners and assume His person and character; have the mind in us which was in Him; enter into his views; imitate him; and be wholly in agreement with him: *But put ye on the Lord Jesus Christ, and make not provision for the flesh* (Romans 13:14). Let us put on decent garments; let us make a different profession; unite with other company; maintain that profession by a suitable conduct, signifying receiving and believing the Gospel.

SPIRITUAL CHILDBIRTH: PHASE 4

SANCTIFICATION

Following the initial acts of regeneration, reconciliation, and justification, where, as a zygote, the "babe" is Christ-formed at the molecular level; brought intimately close to God as a "little child" at the cellular level; declared, as a "young man," to be free of all sin at the organismal level; sanctification finds its fullest manifestation at the populational level, i.e., *Sanctification → Populational → Fetus → Fathers* (refer to Figure 11.1); wherein the *fathers* hold the distinctive position of knowing Him *from the beginning.*

Sanctification is man's input to cultivate the soul for a spiritual hunger and thirst, the impartation into our emotions and personalities the benefits of the fruit of the Spirit. It is the "putting on" of the Lord Jesus Christ; our transformation into the express image of God by the renewing of our mind; the "putting away" of those things that hinder growth in the faith; where God's love, joy, and peace can be established in the deepest recesses of our personality.

Sanctification is a process whereby a person in Christ is *set apart* or *hallowed,* from the Greek (*agiazo*), which stems from *agios,* meaning *holy* or *saint.* The words "sanctify," "sanctification," "saints," "sanctuary," "holy," and "hallowed," are all derived from the same root word (*agios*), as this word and its derivatives appear some 280 times in the New Testament.

In this context, a regenerated, born-again believer, from a beginning zygote to a mature father, is a saint, a*gios,* set *holy* in his sDNA flesh. Nothing may be added to or be taken away, nothing to increase or decrease his purity. He is also sanctified, *agiazo,* set *apart* to God in his nDNA flesh, to continue the process of spiritual regeneration started in his sDNA flesh, until the process of sanctification is complete. Therefore, the whole of the Christian body is both a *set holy* and a *set apart* believer, both *agios* and *agiazo*—set holy in his sDNA flesh, wholly sanctified, where nothing may be added:

Put on therefore, as the elect of God, holy and beloved, bowels of mercies, kindness, humbleness of mind, meekness, longsuffering (Col. 3:12).

When the elect is said to *put on Christ,* he is said to assume all the excellencies and attributes of Christ. Therefore, to "put on" all the *bowels of mercies* listed indicates that Christians are enthroned in our spiritual sDNA flesh with all the excellencies and attributes we need to function in the world.

Not so with nDNA flesh of the Christian body as this entity is but *set apart,* in need of regeneration and sanctification in his nDNA flesh, where much may be added:

And the very God of peace sanctify you wholly; and I pray God your whole spirit and soul and body be preserved blameless unto the coming of our Lord Jesus Christ (1 Th. 5:23).

Sanctification is an ongoing process, initiated immediately upon the instant of regeneration, and is confirmed Biblically to consist of three aspects:

• Positionally
The believer's body, now wholly Christian, is sanctified for the glory of God at the moment of salvation, *set holy* in his sDNA flesh and *set apart* in his nDNA flesh—set apart for holy purposes that are soon to follow.

• Progressively
The believer's nDNA flesh is *being* set apart every moment of his earthly life, adding those things that the Lord directs to enhance his growth and development. In this sense, sanctification is a progress toward sinless perfection, never to be achieved while in the flesh. Sin will remain so long as the flesh continues in its present state:

For I know that in me (that is, in my flesh,) dwelleth no good thing: for to will is present with me; but how to perform that which is good I find not (Rom. 7:18).

The flesh continues to be the great antagonist of man's spiritual nature until death severs the two. It should cause untold joy in the believer that he sins in his body in an inverse proportion to the degree of his sanctification, i.e., the higher the sanctification, the less he will sin; the lower the sanctification, the more he will sin. Strive, therefore, for perfection.

• Perfectively

The believer's sDNA flesh is set holy, obtaining complete separation from the world. This can only happen at the resurrection of the body when not one molecule of nDNA flesh is to be found in the Christian. No more need for progressive sanctification here, no more need for striving after the flesh here:

> Sanctification is in three tenses: *positionally*, the believer is set apart in salvation; it is instantaneous and absolute; *progressively*, he is being set apart all his earthly life; *perfectively*, he is glorified and achieves perfect separation unto God at the resurrection.[1]

Justification and sanctification are inseparable, yet are so distinct. One is a single, legal act, a right to life; the other a continuous, physical act, a change. The one is by righteousness without us; the other is by holiness wrought in us; the one precedes, the other follows; the one is by Christ as a priest, the other is by Him as King; the one annuls sin's damning power, the other proclaims it's reigning power; the one is instantaneous—complete; the other is progressive—perfecting.

Fathers in Christ should be as a beacon shinning in a dark world, to light up the dark crevasses of the soul and spirit. Charles Spurgeon declares:

> Sanctification *begins* in regeneration. The Spirit of God infuses into man the new principle called the spirit, which is a third and higher nature, so that the believing man becomes body, soul, and *spirit*, and in this he is distinct and distinguished from all other men of the race of Adam. This work, which begins in regeneration, is carried on in two ways--by vivification and by mortification; that is, by *giving life* to that which is good, and by *sending death* to that which is evil in the man.[2]

Some of you who ought now to be *fathers* in Christ are still babes, and some of you who now are still *babes* in Christ will be *fathers* by this time tomorrow. In the first, there is very little progress in sanctification; in the second, there is much. Christ has redeemed you by the shedding of blood, purchased you by the sprinkling of that precious ointment, but to little avail if no progress is made; to much if progress is made toward perfection.

POPULATIONAL

As both the Source of Empowering Energy and the Conceptual Information program are manifest at the Molecular, Cellular, and Organismal Levels, the Converter process of regeneration performs its primary function at the Populational Level where we see large groups of Christians assembled together to worship God. It is there that the war between good and evil in each of us is waged, resulting in massive transformations of sinful nDNA flesh to sinless sDNA flesh. We are at liberty to display those characteristic *marks* in the body, traits and observations that are pleasing to Him.

As spiritual growth continues, as sanctification takes root in the early zygote, progressing in the babe, little child, and the young man phases of growth, and now at the populational level, we readily identify the elect of God where we see righteousness and holiness in what we do and in what we say. This is a continual celebration in life of the cure of sin. Our identification is positive. God has determined in His sovereign will to command the helicase, the DNA gatekeeper, to unzip the predestination genes to initiate the process of eternal life, a satisfaction of infinite grace for the death of His Son.

FETUS

On the surface, we see the long warfare between the elect and the non-elect but little do we realize that within the womb is where the war is really fought.

The war began in the zygote stage of growth, between the spiritual and the natural, raged through the blastocyst and embryo stages, and now expands into the final fetus stage with increasing fury. The battle is similar to the conflict between the cells of a mother and the cells of her natural fetus, a battle for the very existence of life, that rages during a lifetime, and ends at death. A defeat at the hands of the enemy would spell doom for the entire human race.

> The spiritual life or nature communicated at regeneration is not the only thing in the Christian: the principle of sin still remains in the soul after the principle of grace as been imparted. Those two principles are at direct variance with each other, engaged in a ceaseless warfare as long as the saint is left in this world.[3]

In the spiritual realm, identical procedures are set in motion. sCells of the new birth fetus are identified by the host nCells as foreign tissue and summons an all-out attack against this foe. The sImmune system senses this intrusion as an attack upon our very nature, and as a violation of the nature of God.

When it all started in the Zygote stage, the sImmunes were out-numbered by the nImmune system and in imminent danger of defeat. A strong nImmune system allows pathogens of every stripe to grow in number thereby weakening the sImmune system. In the embryo stage, however things look a little bit better as the spiritual now begins to grow in its ranks to finally come equal to its enemy's stature. sCells are increasing in number at this stage and is able to trick the host's tissue into believing it is also host tissue and not spiritual fetus tissue in full attack mode.

But not all is quiet and peaceful. The sFetus armies now begin to attack the host! sT-Helper cells are summoned to organize the attack on each nCell. Normally, sNatural Killer cells would move in to destroy an invader but now limit their function to aiding in the transformation of nCells to sCells. sT-Suppressor cells call off the attack once each attack has been completed, adding further protection to the newly formed sCells. In this wise, regeneration never stops—nCells are constantly coming to spiritual life by the power of the Holy Spirit, every moment of every day.

> The miracle of Biblical forgiveness is the theological substrate for the biological immune system of the new birth fetus, to feed upon the mercy and grace of God.

The growth of the new birth fetus is a life-long process of enmity against the flesh to reach full term, with slow ongoing transformations dependent upon close relationships with Christ. The fetus within the new birth is thus forever preserved.

FATHERS
We as Christians can easily be identified in any population as the Body of Christ, for we are not ashamed to display the virtues that identify us; *marks* permanently incised into our body, that are not found in other classes of people, never to be

erased or cleansed. Holiness and virtue are the norm at the Organismal Level and an outward expression of those peaceful virtues into the Populational Level leads to a state of health, both physically and spiritually.

> These four classes constituted the household or family of God; each class, in ascending gradation, seems to have had more light, experience, and holiness than the other. 1. The teknia, beloved children, or infants, are those who are just born into the heavenly family. 2. The paidia, little children, are those who are able to walk and speak; they know their heavenly Father, and can call him by that name. 3. The neaniskoi, young men, are such as are grown up to man's estate; these perform the most difficult part of the labour, and are called to fight the battles of the Lord. 4. The paterev, fathers, are those who are at the foundation of the spiritual family....They are the parents, the father and mother, from whom the family sprang, and who are the governors and directors of the household.[4]

We may glean from our earthly walk with Christ, images of personal behavior. We are seen every Sunday seated in Church worshipping the One who redeemed us; attending Bible study Wednesday evenings; visiting the elderly in nursing homes; offering assistance to patients in the hospital; attending revival meetings and crusades; making yearly pilgrimages to Israel to the Holy Sites; on our knees praying, and singing in the choir.

As light-winged butterflies will not inhabit dark forests and dark-winged butterflies will not inhabit sun-lit fields, elect Christians will not inhabit saloons and taverns nor will the non-elect attend church services. DNA in both groups determines inhabitancy patterns.

To understand why Christians act and behave in this manner is to know all about God, the source and fountain of life as expressed in our sDNA molecules. Moral attributes of a Christian will never be understood by those who are not one of them. Predestination genes, with their divine sequences, will remain hidden within the human genome, never to be fully identified. They will never be isolated, never transcribed, translated, or expressed as proteins by human means alone.

It soon becomes apparent, and more and more evident, that a trust in Christ appears in the consciousness that affords an assurance of salvation; a desire to hear the Word; to pray; a delight to worship; a thirst for Christian fellowship; a desire to be obedient; a longing to tell others, and a compassion for a lost world. These constitute *marks* of the Lord Jesus manifested at the Populational level as expressed in cells of organs.

• Ye have known him
I write unto you, fathers, because ye have known him that is from the beginning (1 John 2:13).

Even mature Christian "fathers" need to be written unto occasionally; nDNA still requires instruction. Remaining natural cells must receive some degree of knowledge of spiritual things prior to the Holy Spirit regenerating; either by looking up into the starry sky at the wonder of God's creation, or understanding their lost condition without God and crying out to Him for more knowledge.

• Ye are Priests
But ye are a chosen generation, a royal priesthood, an holy nation, a peculiar people; that ye should shew forth the praises of him who hath called you out of darkness into his marvelous light (1 Pet. 2:9).

God created a chosen people, a holy nation; there is no capital here, no king here, and whose land is in heaven. Jesus created the elect nation just as He created new matter; as He multiplied the loaves of bread and fishes to feed the five thousand; when He walked on water dispensing the law of gravity; when He stilled the storm; and when He made wine anew.

Christians may lay claim to be a priest after the order of Christ, and not after the order of Aaron, the high priest of Israel. A recent study was done of various segments of the Jewish population who claimed to be descendants of the priestly line of Aaron. A DNA comparison analysis was performed upon the claimants and it was determined that God has sovereignly preserved the priestly line of Aaron to this day!

The human genome contains 44 autosomes and two sex chromosomes. The two sex chromosomes of the female parent are described as XX and that of the male parent as XY. The female parent contributes one X chromosome to her offspring and the male parent may contribute an X *or* a Y chromosome.

Therefore, the child may receive XX chromosomes, a girl, or XY chromosomes, a boy, one chromosome from each parent.

The Y chromosome in the male gender is passed on from father to son intact, and if two individuals in differing generations possess the same Y chromosome, it is proof positive that they are related by heredity.

Therefore, you simply cannot join the Jewish priesthood no more than you can simply join the Christian priesthood. Levites are born into the priesthood because of their telltale Y chromosome inherited at natural nDNA birth; Christians are born into the priesthood because of their telltale Y chromosome inherited at spiritual sDNA birth. The distant parent of the Jewish priesthood is Aaron; the distant parent of the Christian priesthood is Christ!

God is our parent, yes, you! As a new creature in Christ, you are declared to be of a chosen generation, belonging to a royal priesthood, and a kingdom of priests! Not just of any priesthood, but of a *royal* priesthood, sworn into by birthright:

> All believers are unconditionally constituted a "kingdom of priests". . . . The priesthood of the believer is, therefore, a birthright; just as every descendant of Aaron was born to the priesthood. The chief privilege of a priest is access to God. Under law, the high priest only could enter "the holiest of all," and that but once a year, but when Christ died, the veil, type of Christ's human body, was rent, so that now the believer-priests, equally with Christ the High Priest, have access to God in the holiest.[5]

• Ye are the Temple of God
Know ye not that ye are the temple of God, and that the Spirit of God dwelleth in you? If any man defile the temple of God, him shall God destroy; for the temple of God is holy, which temple ye are (1 Cor. 3:16–17).

The method God used for the building of the Wilderness Tabernacle—the *Outer Court*, the *Holy Place*, and the *Holy of Holies*, Figure 21.1[6], was the *pattern* for the Cell in the Human Genome—the *Cytoplasm*, the *Nucleus*, and the *Nucleolus*, Figure 21.2[7]— both bearing a stark relationship to Man—the *Body*, the *Soul*, and the *Spirit*. You are some amazing temple!

No light there. No window to let light in; lit only from the Golden Lampstand, all day, all night, as a reminder that God's presence is always with His people. Man is blind apart from Jesus; man cannot know God apart from Jesus. Who can deny the presence of God in the Holy of Holies of the Tabernacle; in the Nucleolus of the Cell; or in the Spirit of Man?

Figure 21.1: Wilderness Tabernacle

As a royal priesthood, we Christians are authorized to offi-ciate the functions of the Holy Place to our generation. Jesus said: *"I am the way, the truth, and the life"* (John 14:6). The *way* is through sacrifice and atonement on the brazen altar in the Outer Court—in the Cytoplasm, and in the Body; the *truth* is through the workings of the priesthood in the Holy Place— in the *N*ucleus, and in the Soul; and the *life* is wrapped up in God in the Holy of Holies—in the Nucleolus, and in the Spirit.

• **The Outer Court—the Cytoplasm—the Body**
The Outer Court of the Tabernacle finds its biological counterpart in the Cytoplasm of the Cell, and in the Flesh of the human Body; completely surrounding the two inner struc-tures—the Holy Place and the Holies of Holies; symbolizing the Cross where sin was dealt with—our Savior sacrificed on the brazen altar; the altar of sacrifice, where the Israelites

came to worship God and where He now meets with His people. Come now ye people of God and worship the King!

On its eastern side was the Eastern Gate, a representation of Christ as the only way through which you, as worshipper, could draw near to God; always open, never barred; with no one forbidden access who desired to worship. Enter in through the narrow gate where God dwelled, and as you entered, the first object to meet your eyes was the witness of the altar and the need for a blood sacrifice; standing as a monument and a bulwark against those who would seek entrance into the Holy Place; through the Door; without satisfying the righteous demands of the Cross upon his life.

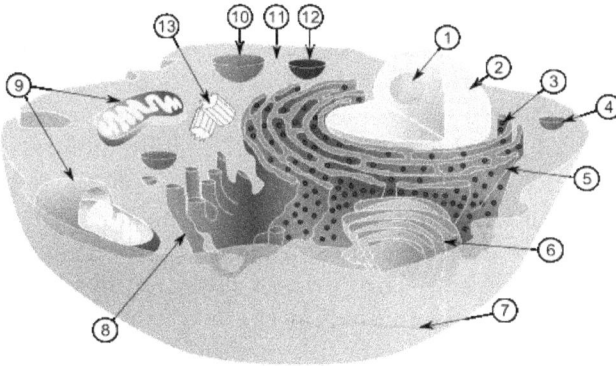

(1)nucleolus (2) nucleus / envelope & pores (3) ribosome (4) vesicle (5) rough ER (6) Golgi apparatus (7) plasma membrane/pores (8) smooth ER (9) mitochondria (10) vacuole (11) cytoplasm (12) lysosome (13) centriole

Figure 21.2: The Human Cell

All the major systems of the Body are here derived from molecular proteins, synthesized in the Nucleolus and transported to the Cytoplasm; to develop pluripotent stem cells of the digestive, endocrine, circulatory, respiratory, urinary and reproductive systems; the factory work floor, where the myriad of things for the spiritual life are put together; assembled into specific organs to function in specific manners; where translation occurs; where mRNA, tRNA, and ribosome subunits come in contact with the endoplasmic reticulum; containing ribosome waiting to enter into union with the regenerated genetic code embedded in mRNA; to produce anticodons and proteins; to produce organs; and finally, to produce—you and me!

• The Holy Place—the Nucleus—the Soul

The Holy Place of the Tabernacle finds its counterpart in the Nucleus of the Cell, and in the Soul of Man where the High Priest officiates—in the heart, and where the Holy Spirit indwells the believer. God will give you a new heart, and a new spirit within your new heart—a new spiritual heart of flesh to cause you to walk in His will, as you are a *royal priesthood.*

A new heart also will I give you, and a new spirit will I put within you: and I will take away the stony heart out of your flesh, and I will give you an heart of flesh. And I will put my spirit within you, and cause you to walk in my statutes (Ezek. 36:26–27).

This is where the various functions of a priest are to be carried out daily; the lighting of the candelabra (Jesus: *I am the Light of the world* (John. 8:12)); the preparation of the shewbread (Jesus: *I am* the *bread of life* (John 6:35)); the burning of incense *(for a sweetsmelling savour* (Ephesians 5:2)). The lampstand of solid gold stood on the left; the table of shewbread on the right; and the golden altar of incense in back before the veil; where only priests could enter and minister; the washing of hands and feet at the laver in the Court before entering through the beautiful curtain—the "door" (Jesus: *I am the door* (John 10:9)); all bear a direct relationship to the Nucleus of the Cell, the seat of the Soul of Man.

Three passageways are allowed through which God may be sought: (1) Through the eastern gate of the Outer Court, through the door of the Holy Place, and through the veil into the Holy of Holies of the Tabernacle where God resides; (2) Through gated pores in the Plasma Membrane of the Cytoplasm, then through nuclear pores in the Nuclear Envelope of the Nucleus of the Cell, to the Nucleolus within; and (3) Through receptors in the skin and through natural openings of the Body, to the human Spirit within. Entrance is guaranteed for those who possess incorruptible seed—enter in; signifying acceptance of the completed act of Christ.

The image of God is marred; entrance is denied for those who reject Christ; unbelievers; those possessing corruptible seed—the antigen/pathogen complex of sin; thus avoiding corruption of holy ground. God finds no delight in these; nor will He ever reside in an unholy Temple, in fallen Man.

Who among us can fathom the beauty and the majesty of the Holy Place, corresponding to the Nucleus of the Cell and to the Soul of Man; where all the major genetic systems assemble together; where emotion, intellect, personality, and will, reside. But, as you move away from God's presence back out into the world, there is only judgment and need of continual cleansing.

• **The Holy of Holies—the Nucleolus—the Spirit**
Personally, I contend the following to be true: That God the Father, the Energizer, resides in the Holy of Holies of the Tabernacle—in the Nucleolus of the Cell and in the Spirit of Man; God the Holy Spirit, the Regenerator, resides in the Holy Place of the Tabernacle—in the Nucleus of the Cell and in the Soul of Man; and God the Son, Jesus, the Informer, resides in the Outer Court of the Tabernacle—in the Cytoplasm of the Cell and in the Body of Man.

A distinguishing feature of the Tabernacle is the initial design, indicating the truth that: As one goes from the outside to the inside—from outside the Outer Court to the Holy of Holies inside; from outside the Cytoplasm to the Nucleolus inside; from outside the Body to the Spirit inside—one goes from the ugliness and despair of sin, to the beauty and holiness of God.

The animal skins of the Tabernacle, covering the Holy Place and Holy of Holies, pointed to the way that leads to God: From the outer covering of Badger skins, displaying the unattractiveness of Jesus to those who reject Him; next, of Ram skins dyed red, symbolizing the shed blood of Jesus; to that of black Goat hairs; and finally to Fine Twined Linen woven together with blue, purple, and scarlet material—the holiness of God; more precious the further you move from the Outer Court inward towards the Holy of Holies, the more beautiful were the materials used; a progression from the less attractive to the more beautiful; from the unattractive badger skin covering on the outside to the final covering of the finest white linen on the inside; the deeper we look into the things of God, the more beauty and splendor we find.

The Holy of Holies in the Tabernacle finds its counterpart in the Nucleolus of the Cell, and in the Spirit of Man: Wherein all the glory and majesty of God resides; in which is found the Ark of the Covenant, the symbol of His presence; upon which is the mercy seat where God speaks to His children; from behind the veil within the Nuclear Envelope; wherein resides the genetics for the creation of the spiritual man; where embedded information resides; where the repository of the incorruptible Word and Life are to be found; the command center where God communes with man by means of cell communications; the Holiest—a small room, 15 feet square, separated from the Holy Place by a veil guarding the entrance through which no man nor angel may touch, let alone enter; except the High Priest in order to sprinkle blood on the mercy seat on the Day of Atonement to confess the sins of the people.

The Holy of Holies houses one piece of redemptive furniture where no artificial or created light may be seen. God's "Shekinah Glory" lights-up the Holiest; no seat there for man to rest; God alone sits on the throne of His glory; the High Priest entering with head bowed, unsandalled feet, and bells; no human voice heard there, only the voice of God.

This is a stark reality of the workings of God sitting on the mercy seat in the Holy of Holies—in the Nucleolus of the Cell, in the Spirit of Man; acting through processes of genetic transcription; ordering and sustaining the genetics of the human body; as in all the starry hosts in the heavens, in all their intricate and minute parts and details.

CHAPTER THOUGHTS

Fathers in Christ are seen to be those who bear the total responsibility of raising-up the younger, more immature, Christians in the faith. They have learned what it means to "put on Christ," to be clothed with Him; to sink into His own garments; to emulate Him; to copy and to imitate Him.

Christian fathers bind together like a bed of roses as seen in the fields, glistening in the cool of the noonday sun, with brilliant red colors as in the bloom of the flowers, emblematic of the way we associate with friends of like precious faith.

This is nothing but the act of sanctification at work in the real world, where, positionally—we have been redeemed by the blood of the Lamb; progressively—of being "set apart" to hear the Word of God; to open our eyes and ears to see and to understand the beauties of the Lord; and perfectly—to await the future redemption of the body where we will see Him, face to face, and hear Him say to us: *Well done, good and faithful servant* (Matthew 25:23).

We are now priests in the holy temple of the human body, offering praise and glory to our awesome God who died for our sins. You now have an obligation, now that your life is hid in God, and bound up in the Holy Place of the Temple—expressed within the Nucleus of the Cell, and in the Soul of Man, to perform the functions of a priest; to let your light so shine in the world; having purified yourself by the washing of hands and feet in the Outer Court; to point the lost world to the only way that leads to God; gaining entrance to the Holy of Holies by means of incorruptible seed.

TIDBITS OF PART VI

- God is Love → The Father → The Source:
  - o The Holy Mutagen, Attacks DNA structure.
- God is Light → The Son → The Informer:
  - o The Light that shines in our heart.
- God is Eternal Life → The Holy Spirit →The Regenerator:
  - o *Except a man be born again.*
  - o The Necessity of–*Ye MUST.*
  - o The Effectual Calling: Drawn, sealed.
- Thoughts → Immaterial, information, intelligence:
  - o Neuropeptides → emotions → feelings.
- The Emotional God—the Fruit of the Spirit.
- FRUIT→sNeuropetide → sEmotions → sImmune Cells:
  - o FAITH → sVasopressin → sFaith → sMonocyte.
  - o MEEKNESS→sAnandamide→sMeekness→sMacrophage.
  - o TEMPERANCE→sOrexin→sTemperance →sDendritic Cells.
- Faith, Meekness, Temperance vs. idolatry, wrath, & revellings.
- Phase 4: *Sanctification →Populational →Fetus →Fathers:*
  - o Sanctification;
    - ◆ Positionally.
    - ◆ Progressively.
    - ◆ Perfectively.
  - o Populational: Christians bind like a bed of red roses;
    - ◆ Fetus: Life-long process of "war" to reach full term.
    - ◆ Fathers: Images of personal behavior.
    - ◆ Ye have known Him.
    - ◆ Seated every Sunday in Church.
    - ◆ Visiting the elderly in nursing homes.

- ♦ Hospital visitation.
- ♦ On our knees praying.
- ♦ Singing in the choir.
- ♦ A desire to hear the Word.
- ♦ A thirst for Christian fellowship.
- ♦ A compassion for a lost world.
- *Put ye on the Lord Jesus Christ!*
- Priests—A *chosen* generation:
  - o A royal priesthood;
    - ♦ Ye are the Temple of God!
- The Outer Court → the Cytoplasm → the Body:
  - o Jesus our Savior!
  - o Enter in though the narrow gate;
    - ♦ Meet Him here.
  - o The brazen Altar;
    - ♦ Blood sacrifice, the Cross.
    - ♦ Where sin is dealt with.
  - o Life → Proteins, organs;
    - ♦ You and me!
- The Holy Place → the Nucleus → the Soul:
  - o The Holy Spirit.
  - o The High Priest officiates.
  - o The Factory work floor, mRNA, tRNA, genetic code.
  - o The Door (*I am the Door*).
  - o Golden Lampstand (*I am the Light*).
  - o Table of Shewbread (*I am the Bread of Life*).
- The Holy of Holies → the Nucleolus → the Spirit:
  - o The Father.
  - o The Ark of the Covenant → His presence.
  - o The Mercy Seat → God speaks from above.
  - o No light there, no seat, no resting, no talking;
    - ♦ Only the "Shekinah Glory" there.
  - o Repository of the incorruptible Word of Life.
  - o All the genetics reside there.
  - o Enter in;
    - ♦ Ye who possess incorruptible seed.
    - ♦ With unsandalled feet.
    - ♦ With bells, head bowed.

PART VII

Eternal Destinies

22

BIOLOGICAL RESURRECTION

"I am the resurrection, and the life: he that believeth in me,
though he were dead, yet shall he live" (John 11:25)

Man is in dire need of a drastic change in his life, of a new birth, to cure him of the sorrowful effects of his initial birth. Man is conceived in sin: Born of a woman; of sinful flesh, fleshy; of no good works; slothful; unpleasing toward God; a renegade; vile; exceedingly sinful; without Christ and without peace in the world; by nature, children of wrath; unruly; disobedient.

> After Adam and Eve's fall, God expelled them from the Garden of Eden and so from the Tree of Life. The theology is that God saw that the worst thing for fallen humanity would be to live forever in a fallen state by having access to the tree of Life. Therefore, it was best that man be driven from the Garden, then given the opportunity to repent and restore his relationship with God, and then die physically with the hope of resurrection to a new, glorified body. Living forever in a fallen body is not salvation. It does not beat death, for fallen man is eternally dead in his heart no matter how long he physically lives.[1]

The promise here is that some day you will change. You will not remain the same as you were all your life. Change is inevitable. Change is already happening in your system—you are not the same as you were yesterday—and you will not be the same tomorrow as you are today. This change that we have been speaking about all along is—regeneration—and it is already occurring in the body of every Christian; cells coming to the knowledge of Christ; waiting for the resurrection event.

To be sure, there are several descriptions concerning the act of resurrection in the Scriptures. Resurrection is a term applied to the elect of God, never to the non-elect. Although the unbelievers will also be made alive, it is not referred to as a resurrection to eternal *life*; but as a resurrection to eternal *damnation*:

> *Marvel not at this: for the hour is coming, in the which all that are in the graves shall hear his voice, And shall come forth; they that have done good, unto the resurrection of life; and they that have done evil, unto the resurrection of damnation* (John 5:28–29).

There is a fine line between regeneration and resurrection. Regeneration and Resurrection are seen as Siamese Twins; inseparably joined together at the hip, and wherever Regeneration goes, Resurrection is sure to follow; and wherever Resurrection goes, Regeneration is not far behind. There can be no regeneration without resurrection, and no resurrection without regeneration. This may seem at first to be paradoxical but to properly apply both terms requires knowledge of the recipient's condition; whether alive or dead at the time, and, most importantly, at what time in history.

MATERIAL OR IMMATERIAL?

Regeneration and resurrection do not imply an immaterial body. Our spiritual body is not some kind of ether or ghost that floats in and out of our natural body, then up and down in the atmosphere. Spiritual is material just as the natural is material; but the spiritual material is a *refined* form of the natural material. The apostle Paul indicated this when the Israelites passed through the Red Sea on their way to Canaan, the promised land, eating and drinking *spiritual* food:

> *Moreover, brethren, I would not that ye should be ignorant, how that all our fathers . . . did all eat the same spiritual meat; And did all drink the same spiritual drink: for they drank of that spiritual Rock that followed them: and that Rock was Christ* (1 Cor. 10:1, 3–4).

While the reference here is to all of Israel (both spiritual and non-spiritual alike) who ate daily of the provisions provided for them, it is nonetheless spiritual food and spiritual drink, where the term "spiritual" is definitely material, not immaterial.

> Here it is understood that in *spiritual rock* and *spiritual drink* there is an intended modification of the idea of the material rock and material drink, but it is not the intent to eliminate the material character of either the rock or drink.[2]

It is not necessary to die and be resurrected before you can assume a spiritual body, to then be able to eat and drink spiritual food. The Scriptures affirm that living Christians are classified as spiritual, those who have been regenerated by the Holy Spirit: *And the disciples were called Christians first in Antioch* (Acts 11:26).

ON EARTH *AND* IN HEAVEN – REALLY?

It would be pure presumption to declare absolutely how God infected the human genome with sin. I know that somehow, in some way, He incised sin *into* the genome. And if He incised sin *into* the genome, then He could excise sin *out of* the genome just as easily.

We also know that the regenerated and resurrected sDNA life is far more glorious than the original Adamic life. He incised into the genome values far more abundant and glorious, values never before seen, nor will ever be completely understood. One of the greatest of blessings afforded to us as Christians is the realization that not only are we living on this planet now, but we are also living in heaven at the same time!

> *If I have told you earthly things, and ye believe not, how shall ye believe, if I tell you of heavenly things? And no man hath ascended up to heaven, but he that came down from heaven, even the Son of man which is in heaven* (John 3:12–13).

No man has ever ascended into heaven *before* the resurrection of Jesus. But after that, I know of two other Biblical persons who are said to hold the distinction of being in two places at the same time. Besides Jesus, they are the apostles John and Paul.

What is required of a person to hold that distinction, of being in heaven and on the earth at the same time? First of all, of absolute necessity, he must be a unique person of two distinct natures. Jesus was the first to hold this position as no one has ever preceded Him into heaven; and who represents a very special case because of His deity and attribute of omnipresence. Matthew Henry comments on the truth that Jesus came *down* from heaven and holds the position as Son of Man and, simultaneously, as of being the Son of God *in* heaven:

> We have here an intimation of Christ's two distinct natures in one person: his divine nature, in which he *came down from heaven*; his human nature, in which he is the *Son of man*; and that union of those two, in that while he is the Son of man yet he is *in heaven*.[3]

We need then to look at His spiritual nature and to draw the conclusion that the proper emphasis is not being placed upon His divine nature but His spiritual nature is in view; residing *both* in heaven and on earth, simultaneously—when He walked on earth, He was also in heaven!

• The Apostle John

It is our position that when a person is regenerated, receiving a spiritual nature to reside with his natural nature, that spiritual flesh is also immediately resurrected. Hence, a Christian exhibits two modes of existence: A fundamental mode on earth while he is yet alive with both n- and sDNA flesh, and the other mode in heaven with sDNA flesh, *fully grown,* not missing one molecule. As the natural nature of man is landlocked in the body, so is the abiding *incomplete spiritual nature.* Therefore, when we walk on earth, we are also in heaven!

Not much is given about the apostle John in the Revelation, except that he was summoned up into heaven, bid to enter through an open door in order to personally view the things that are to be in the future:

> *After this I looked, and, behold, a door was opened in heaven: and the first voice which I heard was as it were*

*of a trumpet talking with me; which said, Come up
hither, and I will shew thee things which must be hereafter. And immediately I was in the spirit: and, behold,
a throne was set in heaven, and one sat on the throne*
(Rev. 4:1–2).

Many expositors have assigned v. 2 to the work of the Holy
Spirit even though *spirit* is written in lower case letters, indicating that spirit could very well allude to John's spirit or spiritual nature. Whatever the case, it is plainly in view here that
John was in heaven with his complete spiritual sDNA body (as
sinful flesh is prohibited), regardless of the mode of his arrival,
which must be related to his transportation by the Holy Spirit.
It was about to be revealed to him the things of the future
from the One Who sat on the throne in heaven, Jesus Christ.

At the same time, it is evident that his earthly body did
not disintegrate or evaporate while he was in heaven. His
earthly body still consisted of nDNA and sDNA flesh as the
spiritual is never separated from the natural except at death.
Therefore, John was simultaneously in heaven with his *complete* spiritual body and on earth with his natural and *incomplete* spiritual body—at the same time. *Awesome!*

• **The Apostle Paul**
The apostle Paul's experience is even more revealing. He
knows that he was in heaven but did not recognize what kind
of body he possessed while there. As with John, that heavenly
body must of necessity be wholly spiritual with no remnant of
the natural. He had an experience of a heavenly matter that
many Christians today will simply not adhere to. Paul writes:

*I knew a man in Christ above fourteen years ago,
(whether in the body, I cannot tell; or whether out of the
body, I cannot tell: God knoweth;) such an one caught
up to the third heaven. And I knew such a man (whether in the body, or out of the body, I cannot tell: God
knoweth;) How that he was caught up into paradise,
and heard unspeakable words, which it is not lawful for
a man to utter* (2 Cor. 12:2–4).

The phrase *I knew a man in Christ* identifies absolutely a
Christian, for no one is *in Christ* except a regenerated Chris-

tian. In addition, he is referred to as a *man*, not as an immaterial ghost or a thing unrecognizable, but as a man.

Another observable fact is that Paul, who is the object of these verses, was not able with any certainty to identify himself as such. He confirmed that he was in the third heaven where it is conceded that the third heaven is the place of habituation of the Triune God. He was in his sDNA body, which he did not, nor could not, recognize at the time: *whether in the body, or out of the body, I cannot tell* (v. 2). Paul confirms his inability to recognize what type of body he was in and relates his experience to that of looking through a dark glass, unable to clearly view his complete spiritual body. It is adhered to by all that the body spoken of in these verses is not any part of the natural, sinful body of a Christian, nor can it be.

Paul yearns, as well as do all Christians, to be totally in his sDNA body, rid of all vestiges of sin, and to be in heaven with his Lord. He knew *where* he was. He did not know in *what* type of body he came with, and heard *unspeakable words,* but was prohibited from telling us all about his adventure.

This is an exact replica of every Christian that has ever lived. Every sDNA body follows the pattern of John and Paul, and especially that of Jesus, though we are not imbued with the attribute of omnipresence. As Paul's body was in heaven and on earth at the very same time, so also our spiritual bodies are in heaven and on earth simultaneously. Unfortunately, our sDNA body on earth cannot be disassociated from our nDNA body until completed, or released at death.

• Flesh and Blood

All those of nDNA flesh are excluded from this experience since our Lord specifically said that: *"no man hath ascended up to heaven,"* that is, no mere mortal man consisting of nDNA flesh and blood will ever ascend into heaven attributed to his own power. This is also confirmed in 1 Corinthians 15:

Now this I say, brethren, that flesh and blood cannot inherit the kingdom of God; neither doth corruption inherit incorruption (v. 50).

This is not saying that there will be no flesh and blood in heaven. If Christians have the blood of Christ flowing through their veins and arteries, the very same sacrificial blood that saved our souls from eternity in hell, the very same blood that we were reborn with as in the regeneration (as the blood of a zygote comes from the father), why prohibit it from residing in heaven? After all, the blood of Christ is incorruptible, is it not?

If flesh and blood cannot inherit the kingdom of God, then how is it that Paul was in heaven with his sDNA? Was there no blood in his body? If no blood in his body, then I maintain that there is no blood in the resurrected and glorified body of Jesus, and that is an absurdity! Christ's blood is precious in the sight of God, infinitely more precious than no blood.

Our resurrected bodies will undergo a complete change. Christ will fashion our lowly bodies so that they will be like his glorious body, complete with spiritual flesh and blood. Every Christian holds this distinction; none shall be excluded.

THE RESURRECTIONS

There is no life in chemical DNA molecules; there is no life in gene expression, from replication down to proteins, only information and codes; but there is life in the cell. Everything comes together in the cell. God does His work in the cell. You can have the work of the Father in the rearrangement of DNA sequencing; you can have the Word of the Son breathing-in the codes of spiritual life; but it will not avail until the Holy Spirit does His work of regeneration and resurrection in the cell. Life happens in the cell, and that cell is immediately resurrected.

The Bible maintains a definite order of occurrence among the living and a definite order of occurrence among the dead. We have followed Scripture by dividing our discussion into two distinct resurrections, in hopes that it will become clearer as to when and where regeneration and resurrection occur in the life and death of both believers and non-believers.

THE FIRST RESURRECTION

There are two indications in the Bible that speak of distinct resurrections. The First Resurrection consists of two parts—the First Part, Part A, of the First Resurrection, occurs at the *future* time of the Rapture of the Church (provided it hasn't already occurred by the time you read this); when all those who are in the grave *awaiting regeneration*, those who *hear* the sound of the trumpet when Christ comes in the clouds; will rise up first to meet the Lord in the air:

> *For the Lord himself shall descend from heaven with a*
> *shout, with the voice of the archangel, and with the*
> *trump of God: and the dead in Christ shall rise first*
> (1Th. 4:16).

Those Christians who are yet *alive,* who have been regenerated by the Spirit, those who hear the same sound of the trumpet, will rise up and join those who preceded them; to meet Him in the clouds; then to be with Him forever. This is the First Part, Part A, of the First Resurrection:

> *Then we which are alive and remain shall be caught up*
> *together with them in the clouds, to meet the Lord in the*
> *air: and so shall we ever be with the Lord* (v. 17).

The Second Part, Part B, of the First Resurrection begins immediately *after* the rapture occurs. Known as the beginning of *the time of Jacob's trouble* (Jeremiah 30:7), or as the *great tribulation* period (Matthew 24:21), it is physically consummated seven years later when Christ comes to the earth again to put an end to the raging Battle of Armageddon; to separate those living into just and unjust categories. In the First Part, He comes in the clouds; this time He comes to earth to stay.

We are mainly concerned with the conditions that exist at Part A, at the time of the Rapture of the Church since you and me, as Christians, whether we are alive now or will die before the time, we will participate in this blessed event.

THE FIRST PART (PART A)
• The Natural Man
Cemeteries are full with two types of buried bodies; those of Christians and those of non-Christians, no other possibility exists. Over against this are unbelievers *now* living but who will die before the time. If they have not accepted Christ as their Lord and Savior, there is no regeneration or resurrection during either of the First Resurrection periods. They have no hope and are forever omitted from a resurrection to eternal life.

Their cause is futile. Their bodies will remain in the ground until the time of the Second Resurrection; not as a glorious resurrection occurring as at the Rapture of the Church for Christians, or at the Second Part, Part B, of the First Resurrection when He comes to the earth; but a resurrection of damnation occurring at the end of the 1,000 year Millennial Kingdom age; to await judgment at the Great White Throne Judgment Seat of Christ; a life of eternal damnation as their names are not to be found written in the Lamb's Book of Life:

And whosoever was not found written in the book of life was cast into the lake of fire (Rev. 20:15).

• The Spiritual Man

In natural life, not all cells are regenerated to life at the same instant. After the initial, non-experiential event, it becomes a continuous, experiential, ongoing process in time. The new birth follows the pattern of the old birth, where it starts out in life as a one-celled zygote, and continuously adds new cells, cell by cell, over a period of time until it finally births. It is an ongoing process of sanctification.

Upon the earthly death of the Christian body prior to the rapture, the sDNA portion immediately goes to be with the Lord in heaven (reunites with his complete spiritual body already resurrected and seated), for he cannot sin and therefore, is not subject to death, burial, or decay. The not-yet regenerated nDNA portion of his body is placed in the ground, less sDNA flesh, to await its regeneration and resurrection events.

These buried bodies are not the same as the non-elect nDNA bodies in the ground and differ immensely. Their dust is precious in the sight of God and are said to be *dead in Christ*. They are the bodies who died physically but have not yet been called-out by the Holy Spirit unto regeneration. These bodies are nonetheless sanctified, or set apart, from all the rest of those who are in the ground, of those who *cannot hear*.

For those of us who are alive, going about our daily business, the Christian will experience, at the time of the rapture (as will also the dead in Christ), firstly, the total regeneration of the natural nDNA portion of our earthly bodies by the Holy Spirit, and secondly, now totally regenerated, will immediately be resurrected in the same time frame: *In a moment, in the twinkling of an eye, at the last trump* (1 Corinthians 15:52).

This raptured sDNA body will unite with his sDNA body already resurrected in heaven when he first believed. We justly maintain that, at the moment of regeneration, the new birth actually ushers in a *spiritual, biological resurrection.*

> *Even when we were dead in sins, hath quickened us together with Christ, (by grace are ye saved;) And hath raised us up together, and made us sit together in heavenly places in Christ Jesus* (Eph. 2: 5–6).

All four of the verbs used are in the Greek *past tense indicating completed acts.* The Scriptures remain true: At the moment of regeneration, He made us alive *(hath quickened us)*; we were resurrected *(raised us up)*; and now are seated *(made us sit together)* with Christ. But why are we living here now, in the present? Shouldn't we be in heaven as the passage indicates and not on earth as we are? Allow me at this point to restate what we've been saying all along: A Christian is seated in heaven with his complete sDNA body residing with Christ, *as well as on the earth* with his incomplete sDNA body residing within his nDNA body. Without these assertions, it would be very difficult to ascertain the truth in this matter.

THE SECOND PART (PART B)

Going forward, the time period that incorporates the Second Part, Part B, of the First Resurrection, starts the moment that the Rapture in the First Part is completed, where all Christians are with the Lord in glory; where all dead unbelievers that remain in their graves, and those still living, are in for a great shock in the future Second Resurrection.

• The Natural Man

At the start of this time period, humanity will be void of all living Christians. If you are unfortunate not to have participated in the rapture event, unbelievers who will die during this period will be placed in the ground to join those who died without Christ in Part A, to await future events.

Some living unbelievers in this time period will have the experience of living through this seven-year Tribulation period. They will participate in the judgments pronounced upon them when Christ comes to the earth to bring to a close gentile rule; but there will be no escape—they will die in unbelief and placed in the ground.

• The Spiritual Man

Believers in this period who have died *prior* to the coming of Christ will be resurrected; those who are still living (Jew and Gentile alike) will be totally regenerated, but *not* resurrected; and will advance *into* the Millennial Kingdom on earth to live 1,000 years under the rule of Christ. This last event starts the time period of the Second Resurrection.

> *But the rest of the dead lived not again until the thousand years were finished. This is the first resurrection. Blessed and holy is he that hath part in the first resurrection* (Rev. 20:5–6).

THE SECOND RESURRECTION

The end of the tribulation starts the time period of the Second Resurrection and culminates at the end of His 1,000 year millennial reign on the earth. Believing men and women will live a life as God originally desired them to live. They will have children that could grow up for 1,000 years under the direction of King Jesus and will have all the resurrected saints of all ages as personal, spiritual mentors. But there will still remain one problem: sin has not yet been totally eradicated.

• The Natural Man

Even though the children of the millennium will have the advantage of the Lord's divine presence, and to behold the evidence of eternal life gifted to the saints of all ages, many will not repent!

Here we plainly see the *fallacy* of the Deceitful Error family espousing "free will." Even though they will behold Him daily, they will refuse His gracious gift of life. The impending judgment and sentencing will be swift and complete. The Lord will assign the now *living* unbelievers, and all the *deceased* unbelievers of all ages *now* resurrected, to their eternal doom. This is the finality of the Second Resurrection, or second death of the unjust. Believers in Christ are nowhere to be seen here:

> *But the fearful, and unbelieving, and the abominable, and murderers, and whoremongers, and sorcerers, and idolaters, and all liars, shall have their part in the lake which burneth with fire and brimstone: which is the second death* (Rev. 21:8).

• The Spiritual Man

God will present to new-born, regenerated believers, born during the millennium, with perfected bodies, to live in that state for all eternity, to gaze upon the face of Him who died for their sins. Satan and the manifestation of sin, with all his deceitful "works of the flesh," will finally be eradicated, never to be seen again!

CHAPTER THOUGHTS

It is here acknowledged that other small groups of the just have participated in the first resurrection throughout the ages of time that have not been mentioned. Although Elijah and Enoch cannot strictly be classified as a resurrection, as neither of them suffered death: *Elijah went up by a whirlwind into heaven* (2 Kings 2:11); and *Enoch walked with God: and he was not; for God took him* (Genesis 5:24); and a group of O.T. saints followed Jesus in His resurrection:

> *And the graves were opened; and many bodies of the saints which slept arose, And came out of the graves after his resurrection* (Matt. 27:52–53).

The penalty of sin is infinite in the sight of a holy and righteous God, necessitating a Being with an infinite supply of energy and power—Omnipotent, to address the genome and affect the only cure possible; a Being of infinite capacity of conceptual thought—Omniscient, to direct and order the steps of His creation; and a Being to infuse an endless life—Omnipresent, into an otherwise dead flesh.

The making of a Christian is the same for everyone that is born of the Spirit, for at one time in his life he was like one of them, unruly and disobedient: but now there is no life apart from God; no life worth living without Him; and no goal won if not to glory in Him.

The Father predestinates, elects you, knows you eternally, loves you; the Son shines in your heart, illumines your mind, lights your path; and the Holy Spirit regenerates, quickens you, resurrects you—a raising from corruption to incorruption, from morality to immorality.

Grab hold of the truth of the Scriptures; turn neither to the right nor to the left; be constant in Him; *wait for His coming!*

23

THE NEW BIRTH:
THE SONS OF GOD

"Put on Christ" (Galatians 3:27)

Whether it is through death or through the Rapture of the Church, when Jesus comes in the clouds to receive His own, each human body indwelt with the Holy Spirit will experience the total resurrection of our natural body, into a glorious and redeemed new spiritual body; a body suitable for the glorified soul and spirit in heaven, awaiting the reunion of the holistic person.

We started our earthly, natural life very human, was in time blessed with spiritual life, but we never changed our basic humanity. We will remain human for all eternity, but it will be a *refined* humanity by passing through the raging fires of metamorphosis. The incomplete Christian, that portion of the body not redeemed at the time of death, God will complete the transformation, where the entire body in the grave will be redeemed. There will not be one atom, one molecule, one protein, or one single cell where it will not receive the life-giving power of the Holy Spirit.

So also is the resurrection of the dead. It is sown in corruption; it is raised in incorruption: It s sown in dishonor; it is raised in glory: it is sown in weakness; it is raised in power: It is sown a natural body; it is raised a spiritual body. There is a natural body, and there is a spiritual body. . . . As is the earthy, such are they also that are earthy; and as is the heavenly, such are they

*also that are heavenly. And as we have borne the image
of the earthy, we shall also bear the image of the heavenly* (1 Cor. 15: 42–44, 48–49).

SPIRITUAL CHILDBIRTH: PHASE 5

GLORIFICATION

*Moreover whom he did predestinate, them he also called:
and whom he called, them he also justified: and whom
he justified, them he also glorified* (Rom. 8:30).

Glorification is the fifth and final step in the chain of
events in the life of a believer: *Glorification → Adoption →
New Birth → Sons.* If you are able to read this page, then you
are not yet glorified, for glorification of the redeemed occurs
when Christ returns to earth. When?

*In a moment, in the twinkling of an eye, at the last
trump: for the trumpet shall sound, and the dead shall
be raised incorruptible, and we shall be changed. For
this corruptible must put on incorruption, and this mortal must put on immortality* (1 Cor. 15:52–53).

From the preceding passage, it is clear that glorification
ushers in a total change *to* incorruption and immortality, *from*
a state of corruption and mortality, a "putting-on of Christ." A
Christian's glorification describes his ultimate and complete
conformity to the image of Jesus Christ. God makes us like His
Son. It is the final link in the great golden chain of salvation,
reaching from eternity past to eternity future. To be glorified
is another way of saying that the believer is "conformed" to the
character and image of Christ; which is God's ultimate purpose for the Christian; a long awaiting and a yearning to unite
with the previously departed soul resident in heaven.

Glorification is the final step in the application of redemption. It will happen when Christ returns and raises from
the dead the bodies of all believers for all time who have
died, and reunites them with their souls, and changes the
bodies of all believers who remain alive, thereby giving all
believers at the same time perfect resurrection bodies like
His own.[1]

The glorified resurrection body is a spiritual body, not a spirit body. Our body will not resemble a ghost or some invisible or intangible thing floating around in some sort of ether. We will be able to recognize relatives and friends and they will recognize us. God is able to take our molecules from the dust of the ground, complete with our nDNA, and combine them to form our new bodies; and we will see Christ with nail prints in His hands and feet, and a spear wound in His side as Thomas witnessed. We will have substance just as we have substance now but it will be a *refined* substance, that which is suitable for the conditions in heaven. Christ left nothing of His material body in the tomb, and Christ will gather the elements of our body and quicken them into the likeness of His glorious body:

> It is not wise for us to attempt to say much as to when or how the spiritual body comes. We know that it shall be the fitting garb of a ransomed and glorified spirit. We know that it shall be itself a pledge and trophy that of all Christ got from the Father He has lost nothing. It shall represent the dust redeemed, the body ransomed from the grave. How it is woven in the hidden secret of the life after death, we may not venture to surmise. If we have watched how the body, even here, puts on a likeness and correspondence to the real man, to the life within, it will not be difficult to think that for the ripening Christian his future body is being prepared by the Spirit of Christ dwelling already in this mortal frame, and quickening within it that which is to live for ever.[2]

There is something about the human body that God adores, for He will not leave our bodies in the grave to return to the dust of the ground, for it is predestined to be raised an incorruptible body after the likeness of His body. We lose two characteristics of humanity when we participate in the resurrection of the body and in its glorification—we lose our natural neuropeptides and our natural immune system! We won't need them where we are going.

ADOTPION

And not only they, but ourselves also, which have the firstfruits of the Spirit, even we ourselves groan within ourselves, waiting for the adoption, to wit, the redemption of our body (Rom. 8:23).

The apostle Paul links the placing of the mature sons of God with the resurrection of our bodies. Adoption never happens during our earthly life. All we can do is to *wait for the adoption, to wit, the redemption of our body*. There is to be a final birthing process. The body is not fully positioned as adopted until we are with the Lord in the resurrection!

The placing of an individual into the family and household of God establishes a legal relationship over an otherwise unrelated child, but imparts no parental nature or inheritance. It is the resurrected family of God that is composed of the actual and legitimate offspring of God, imbued with eternal life.

> Adoption (*huiothesia*, "placing as a son") is not so much a word of relationship as of position. The believer's relation to God as a child results from the new birth whereas adoption is the act of God whereby one already a child is, through redemption from the law, placed in the position of an adult son. The indwelling Spirit gives the realization of this in the believer's present experience but the full manifestation of the believer's sonship awaits the resurrection, change, and translation of saints, which is called "the redemption of the body.[3]

NEW BIRTH
And as we have borne the image of the earthy, we shall also bear the image of the heavenly (1 Cor. 15:49).

The new birth has its final fulfillment at the final resurrection. During our natural life, a new birth happened every time a new cell comes in direct contact with the Holy Spirit, and is thus fully regenerated, but the *total* regeneration awaits the finality of the resurrection, a glorious one-time event; to bear or to *wear the image* of the heavenly (How the earthly image reunites with the heavenly image, I do not know).

No one knows what a spiritual sDNA body looks like, but it must be such as to be equal in composition to Christ's resurrection body, for we shall be like His glorious body. We will be able to go through doors as He did in the upper room; to appear or disappear at will; to eat, to talk, to walk; and to rise up into the air and disappear into the upper atmosphere as He did. It has no need for sleep, never tires, will never grow weary, nor will ever have need of hospice care.

The Body of Christ, the Church triumphant, will live forever. We will have the very same sDNA as Christ has in His spiritual body, for we shall be like Him. Our sDNA will be His sDNA, faithfully copied through progressive gene expressions until the resurrection when it will be complete. We will then sever all connection with our nDNA parents and will assume total connection to our divine parents.

Our soul and glorified spirit await the reunion of the holistic person, for it will not be until we receive our resurrection body that we will visually see the full impact of all the physical and spiritual changes that have taken place in time. As He walked the earth during His ministry, His divine Glory was hidden from our eyes; so also the glory of the elect is now hidden from our view, waiting to be revealed. Our sDNA will be fully metastasized in every cell, a consequence of advancing from a state of progressive sanctification in time to a final and perfect sanctification in eternity. God will then annul the Second Law of Thermodynamics, the law of disease and decay; make final adjustments to our nucleotide structures; breathe in His final Conceptual Thoughts to our code of life, to be totally acclimated to an endless existence in heaven.

It will be a redeemed body; impeccable, unable to sin; sustained by the Divine Presence; glorious and powerful and incorruptible; honorable and spiritual; bearing the image of the heavenly. We will witness in each other a taste of the image and likeness of our awesome God; in what He has accomplished in the ripening of the lives of the redeemed; molded in time by the Holy Spirit, and presented to Christ as His Bride.

What is the nature of the resurrected body? *How are the dead raised up? and with what body do they come?* (1 Corinthians 15:35). This verse addresses only those who are dead in Christ, the main point being that the body must first die. We have shown that when God regenerates a person, He penetrates the plasma membrane of a natural cell and causes a transformation or a metamorphosis to occur; changing the cell to a spiritual cell. However, what we do not recognize is that something has died! The natural cell has died!

We miss this altogether. You cannot have regeneration without the cell dying. Moreover, you cannot have resurrection without raising that which has died and been previously regenerated. Regeneration requires death; and resurrection requires regeneration. The natural body must first die. There

will not be a solitary person in heaven that has not previously
experienced death once.

Death of cells may come in different ways to different people. The non-elect, no matter during what period in life, will
die one way or another. They will then be "made alive" and
resurrected to an eternal death (not as a Christian is regenerated to eternal life).

The elect that have died in this life, whose sDNA is with
the Lord, but with the remaining nDNA still in the ground
awaiting this event, will undergo a final regeneration to full
sDNA status, producing a death blow to all the remaining
natural cells; and secondly, to perform an immediate resurrection on all the newly regenerated cells to a life in Jesus.

SONS

*Behold, what manner of love the Father hath bestowed
upon us, that we should be called the sons of God* (1
John 3:1).

Each of us was physically born in trespasses and sin; regenerated and conceived spiritually as a zygote—a "babe" in
Christ—at the molecular level by the Holy Spirit; reconciled as
a blastocyst—a "little child" of God—at the cellular level; justified as an embryo—a "young man of valor"—at the organismal
level; sanctified as a fetus—a "father" in the faith—at the
populational level; and now, finally, after what seems like a
lifetime of living, we arrive at where we all were born to be in
the first place—glorified in the new birth—as "sons of God"—
at the level of *adoption, to wit, the redemption of our body*
(Romans 8:23).

The final expectation of becoming sons of God, as is also
the condition in the new birth, must still await the finality of
the resurrection, wherein the total body is composed of spiritual flesh, and *adoption has been conferred* by the Father. The
Christian on earth is not a son in the strict sense of the word
since the Christian body consists of two indivisible parts, natural and spiritual. It is only the totally spiritual Christian,
with no part of the natural remaining, which may enter into
the final phase of adoption by God.

We know that our transition to heaven will result in a
change in our spiritual nature, but the change will not affect
the continued accumulation of the marks of a Christian in our

new existence. It does not end when one has been glorified with a life in the presence of the Lord in heaven. Christians in glory will continually add to their spiritual growth in new areas of endeavor, but now not revealed to us in any great detail. But part of what is revealed to us is very revealing:

ADMITTANCE TO HEAVEN
• **Who will be there?**
They which are written in the Lamb's book of life (Rev. 21:27).

There is *the* book of life and *they which are written* in it are all those who have been made righteous by the shed blood of Christ; wherein we will have free access to the Lamb, enjoying new, fresh, and eternal encounters with the Lord.

But then there are *other* books in God's library as revealed in Revelation 20:12: *And the books were opened: . . . and the dead were judged out of those things which were written in the books, according to their works.* Those not found in the Lamb's book of life will not be found in the New Jerusalem.

• **When shall we be there?**
In a moment, in the twinkling of an eye, at the last trump: for the trumpet shall sound, and the dead shall be raised incorruptible (1 Cor. 15:52).

When? In a moment (Gk, *atomos*), in an "atom of time," *in the twinkling of an eye*, those in Christ will hear the sound of the *trump* from the trumpeter; no matter if some of us are in the grave or perhaps still alive, from the foundations of the world until now; from one extremity to the other; we will respond to that blessed sound and rise up to meet the Lord in the air on that great and awesome day; to see Him as He is, for we shall be with Him forever and ever.

• **With what body do we come?**
But God giveth it a body as it hath pleased him, and to every seed his own body (1 Cor. 15:38).

Every Christian will have his or her own body as God has determined it to be. The body will be totally spiritual in that it will be eternal, incorruptible, immortal, and *fully clothed*, but will differ in glory and in power from one to another. Man or nature itself, in all their wisdom and splendor, neither can affect this miraculous act, for what we see here in the new creation from death to life, is the almighty power of God. He manages the whole of the work, that every seed shall have its own glorious body with no defect; as the wheat germ shall never produce barley, or the rye ever to bear oats.

HEAVEN PROPER
• New Heaven, New Earth—No More Sea
And I saw a new heaven and a new earth: for the first heaven and the first earth were passed away; and there was no more sea (Rev. 21:1).

The world that we know is not permanent. The apostle Peter tells us that it will all burn-up some day in a fervent heat. How shortsighted it is to endeavor to build a kingdom in this world. It's like building a sandcastle on the beach at the seashore. The waves come in and the waves go out, and so does the sandcastle. A wise person will look for a new heaven and a new earth, more substantive and real than what we may be experiencing now.

We oft-times hear the question: "What do you think it's like in heaven?" The Bible says more about hell than it does about heaven. We know *of* heaven but we know very little *about* heaven. Even our hymnals tell us more about heaven than the Bible tells us. We thus are likely to have more wrong ideas about heaven than what is actually revealed.

Though the Bible does speak about heaven, it is not possible to determine definitively its character and nature since it is the dwelling place of God, which, of necessity, will remain eternally hidden from our senses. One thing we do know, however, is that heaven will be vastly different from our current physical universe. We know not what the extra dimensions will be like, but heaven will be a multi-dimensional universe, supporting a multi-activity existence, with multi-spiritual endeavors.

God finished creating on the sixth day, rested on the seventh as stated in Genesis, and some day when He resumes His creative acts, God will again create—a new heaven, a new earth, and a New Jerusalem. It seems certain that the new universe will have different physical laws. A new heaven and a new earth gives credence to the fact that the existing heaven and earth have been drastically modified (by the introduction of sin) and are not acceptable for habitation in the age to come, therefore they are said to have "passed away." It is interesting to note that the Bible says "a new earth and a new heaven," not "another earth and another heaven."

The Greek for "new, *kainos*" has the nuance of "new, especially in freshness, unused or unworn", which could render the passage quite differently. In this wise, there is to come a "fresh earth and a fresh heaven" that will bear a direct correspondence to the present earth and the present heaven. Perhaps it all has to do with the present likeness and the present correspondence that the old creation (the old man) has with the new creation (the new man).

In any event, it will be a glorious new and fresh earth, and a new and fresh heaven, as we ourselves are new, *kainos,* creations. It is as if God took pieces of hard black coal and soft coke and placed them into a roaring furnace to liquefy and refine them, to remake them anew or fresh, into a 24-karat diamond. It is still the same material, but is now entirely different.

It is not probable that the sea will be done away with (I think). In the Scriptures, sea is oftentimes associated with chaos and evil, the epitome of which is the beast, "the antichrist" of Revelation 13:1: *And I stood upon the sand of the sea, and saw a beast rise up out of the sea.* Perhaps the Bible is simply stating that none of the evils that befall us now will be available in the new heaven to tempt the saints. Also, if there is to be no more sea, then many of our physical laws will be suspended. Evaporation of seawater into the air to form clouds would be nonexistent, meaning that it probably would never rain again upon the earth. Many speculations that are more exotic could be cited, but personally, I love the sea and would severely miss fishing for a tasty morsel of fried catfish.

• New Jerusalem—Our new home
And I John saw the holy city, new Jerusalem, coming down from God out of heaven, prepared *as a bride adorned for her husband* (Rev. 21:2).

We have here but one verse for the new heaven and new earth, but twenty-two verses in the Revelation for the New Jerusalem. The city is a real city just as the universe is a real

universe. Some expositors would have us believe the opposite
is true; that the New Jerusalem is not a real city and therefore
take it to mean a spiritualized designation, not to be taken as
literal. But the New Jerusalem is to be understood as literal
and is to be our new home, the dwelling place throughout
eternity for the saints of all ages. This fulfills the hope of
Abraham for a heavenly city, *prepared as a bride adorned for
her husband* (v. 2), where intimate union is to be enjoyed with
our Savior, Jesus Christ. It pictures how beautifully will be
our relationship with Him, united for life, like a wedding ceremony between two lovers. Angels are not said to be living
here. That is the picture we are impressed with; spiritually
married to Christ—our husband! And we—His bride!

• The dimensions of it
*And the city lieth foursquare, and the length is as large
as the breadth: and he measured the city with the reed,
twelve thousand furlongs. The length and the breadth
and the height of it are equal* (Rev. 21:16).

Twelve thousand furlongs is equal to 1,500 miles on its
length, width, and height, a large cube of 7,920,000 feet for
each dimension. Assuming 30% of the ground floor is dedicated
for streets, roadways, recreational areas, forests and fields,
and the like (no seas), and if we assign each person a one acre
living space with a 10,000 square feet mansion with 20 ft. ceil-
ings, the New Jerusalem would comfortably house up to 1 billion people! How many upper living levels there are, I don't
know. What a glorious future we will have; a fresh new heaven
and earth, and a New Jerusalem, to enjoy it all.

• The great wall, foundations, and names
*And the wall of the city had twelve foundations, and in
them the names of the twelve apostles of the Lamb. . . .
And the building of the wall of it was of jasper: and the
city was pure gold, like unto clear glass. And the foundations of the wall of the city were garnished with all
manner of precious stones. The first foundation was
jasper; . . . the twelfth, an amethyst. . . . and the street of
the city was pure gold, as it were transparent glass* (Rev.
21:14, 18–21).

The Apostle now turns his attention upon the city itself and speaks about its materials and construction, in describing its walls and gates. This is a real city but we need to understand that it is a vision given to the apostle. He sees a real city where the saints are to dwell forever, but as he describes it, keep in mind also of the analogy of the city to the bride and the bridegroom, between you or me, and Christ.

The city is to have a wall. But why a wall? What need is there for walls in the New Jerusalem? There are no enemies to attack, no wild animals to devour, no one that we would want to leave out, or even to leave in. So, why the wall? One explanation is that the wall will eternally signify the complete safety of its citizens, that the city will stand forever, and cannot be shaken nor fail; an eternal city.

The city has a wall with twelve gates; each inscribed with the name of one the Twelve Tribes of Israel, each side having three gates of entrance. The wall around the city has twelve foundations, one atop the other, each named for one of the Twelve Apostles of the Lamb of God, signifying that both the O.T. and the N.T. saints may live together in the New Jerusalem. The materials used in its construction portray a wall and a city of immense beauty; starting with the first foundation of jasper, a sparkling deep green stone, and ending with the twelfth, a purple amethyst, portraying the beauty and the magnificence of the city of God; each jewel, sparkling with a different color, shinning as the splendor and glory of God. What a place to live in forever: *And the street of the city was pure gold, as it were transparent glass*— and what a bride we are to be in the eyes of our Lord!

• No more night, candles, or sun
And there shall be no night there; and they need no candle, neither light of the sun; for the Lord God giveth them light (Rev. 22:5).

The Glory of God illumines the City of God and the lamp is the Lamb of God, the Light of the World! It is a city of purity, each and every corner of it radiated with the brilliance of the glory of God; no need for the sun to light at day; no need for the moon to light at night. Night represents danger, terror, and crime and these are seen to pass away as the Lord is the lamp that enlightens all things, bringing everything into view.

It represents everything that is opposed to God—unbelief, rebellion, and ungodliness. Even these will be no more, unable to penetrate the brilliance of His glory. But the righteous *shall walk in the light of it* and be glad, for *the gates of it shall not be shut* (Revelation 21:24–25).

- **No more tears, death, sorrow, crying, or pain**
And God shall wipe away all tears from their eyes; and there shall be no more death, neither sorrow, nor crying, neither shall there be any more pain: for the former things are passed away (Rev. 21:4).

Tears represent all misfortunes, agonies, and tragedies that this life has to offer. Those listed are the negatives of life and surely they will "pass away," for God is not some distant deity ruling in some remote corner of the universe. He is intimately involved with our lives and is concerned with every facet of our new existence: *And God shall wipe away all tears,* to include all that is associated; tears will be replaced with gladness; death with life; sorrow with joy; crying with laughter; and pain with pleasure—for they shall all pass away.

But what about the positives—are there not any? We shall be like Him, but in what sense are we to be like Him? What are the positives? Can we truly understand the positives if they were enumerated to us? I strongly believe that we do not have the experiential capacity to understand the positives that we will enjoy, that is to say, our inheritance, for *it doth not yet appear what we shall be* (1 John 3:2), or shall do, or shall say.

All this sounds inconceivable to us, unable to understand, for we have never seen anything like it; never heard anything like it; nor could never understand anything like it; for we have nothing to compare it to in this life. Everything will be renovated, renewed, and refreshed. Every experience in eternity that comes our way will always be fresh, never static, never mundane; then, to pass on to the next exciting event.

- **The Temple**
And I saw no temple therein: for the Lord God Almighty and the Lamb are the temple of it (Rev. 21:22).

God dwells there in the city with His people, but there is no temple. In the O.T. temple there were several means of access that were restricted to the worshipers. It was open, unrestricted access to the Court of the Gentiles where any nationality could enter, but restricted to Israelite men and women to the Court of the Women. The Court of the Israelites was exclusively limited to male Israelites, and just for priests who performed daily sacrifices in the Court of the Priests. The Holy Place was restricted to the working priests, and the Holy of Holies just for the High Priest only once a year; carrying the blood of the sacrifice to observe the Day of Atonement for all of Israel; entering with fear and trembling to sprinkle the blood upon the mercy seat.

But here, there is no temple and no restrictions, for sin is atoned for and removed. The city is cleansed forever. There will be direct, immediate access into the presence of God. We will live in His presence, within His glory, the unimaginable glory of the City of God and of His bride.

• The River, Tree of Life, and Twelve Fruit

And he shewed me a pure river of water of life, clear as crystal, proceeding out of the throne of God and of the Lamb. In the midst of the street of it, and on either side of the river, was there the tree of life, which bare twelve manner of fruits, and yielded her fruit every month: and the leaves of the tree were for the healing of the nations (Rev. 22:1–2).

The apostle John now shows us the imagery of a pure river, crystal clear; a river *of water of life*, which bears resemblance and allusions to the river flowing out from the Garden of Eden; parting into four riverheads and from there, furnishing life-sustaining waters to all areas of the Garden. This river is seen as flowing down from *the throne of God*, the source of all life; down through *the midst of the street* of the city; and with *the tree of life* on each side of the river, bearing *twelve manners of fruits*. The tree yields its fruit every month *and the leaves of the tree were for the healing of the nations*.

The imagery of the tree of life healing the nations bears no relationship to physical or mental healing in the city, as there is no need for this type of healing. The leaves will be an eternal reminder of the life-long saga of the life and death experienced in the present earth, a reminder of all the blessings to come; of constant eternal health; and of a deeply satisfying life afforded to all in our new home; never a want or a need throughout the entire city. The environment will be perfect, spiritually pure, and abundant in all the things that God has in store to lavish upon His children.

LIFE IN HEAVEN

• **Who are we?**
Come hither, I will shew thee the bride, the Lamb's wife (Rev. 21:9).

Our relationship to Christ is described as the glory of the New Jerusalem coming down from God out of heaven. In symbolic terms, the apostle John relates the bride as the Lamb's wife, that is, the wife of Christ. He saw the real city of Jerusalem coming down, but it was described to him in terms of what he was accustomed to seeing and experiencing on the present earth, as there is no language that can adequately describe the relationship between the bride and the bridegroom; between the Church and its head, Jesus Christ.

Currently we are *the bride in waiting*, and He is the bridegroom. In the future, God will consummate all things, and all the saints in glory will be married to Jesus. He will be our husband and we will be His faithful wife.

• **He is our what?**
I will be his God, and he shall be my son (Rev 21:7).

This is the language as used in the Abrahamic Covenant, way back in the early chapters of Genesis; in the language used in the Davidic and in the New Covenants. He is our God and we are His people. That is the Biblical covenantal story of history, the climatic promise of covenants in that God binds Himself to us, His sons. And here we have it listed again in the final pages of the Revelation of Jesus Christ, all the promises of God made in past history descending upon man. Not only will He dwell with us as His people, but He will also be with us and be our God (v. 3). Who is it that can fill in all the details that these verses allude to? What a magnificent thing it is to think about heaven, so much joy, eternal joy, with Jesus, our Immanuel—God with us, a relationship never before equaled on earth or in heaven. If the promise of God *to be with us* is not true, then heaven itself, with all its magnificence and beauty, would be like hell itself.

• Will we see Him? Know His name?
And they shall see his face; and his name shall be in their foreheads (Rev. 22:4).

A great blessing ever afforded to man—we *shall see His face,* the face of Christ, the Lamb of God. How precious the moment. Moses was never allowed to see the face of God but only permitted to see His backside as He passed by: *And I will take away mine hand, and thou shalt see my back parts: but my face shall not be seen* (Exodus 33:23), *for there shall no man see me, and live* (v. 20). And the prophet Isaiah was only permitted to see the "train" or the hem of His garment.

> *In the year that king Uzziah died I saw also the Lord sitting upon a throne, high and lifted up, and his train filled the temple. . . . Woe is me! for I am undone; because I am a man of unclean lips, and I dwell in the midst of a people of unclean lips: for mine eyes have seen the King, the LORD of hosts* (Isa. 6:1, 5).

Just the sight of the glory of God's back as He passed by or of the hem of His garment, brings great men of the faith to their knees in total humility and awe. But we shall see Him as He is, face to face, and this will have a transforming effect on all of us, just as Moses had when he descended from the mount—*the skin of his face shone* (Exodus 34:29); or of Isaiah, his *lips* were touched by a *live coal* from the alter, was made pure, and God sent him to be one of the greatest prophets in all of Israel (Isaiah 6:7). The very sight of Jesus will have an effect upon us that we have never imagined.

We will be constantly changing throughout all eternity, forever being changed into His image; always growing in grace. We will love our fellow inhabitants, not as we loved prior to glory, as if this kind of love is to be continually sought after, but we will love as God loves His Bride, the Church triumphant. We will not love according to "political correctness" as we sometimes do in this life, to the complete abandonment of our morals, but will love totally, sincerely, and spiritually.

• **Will we serve and reign with Him?**
*His servants shall serve him: . . . and they shall reign
for ever and ever* (Rev. 22:3, 5).

God and the Lamb shall be in it. It will not be a place of
idle pleasure, where everyone will have 70 or so virgins to lavish their affections upon, as some religions would have it to
occupy the time. It will be a life of service and activity as the
Lord may direct. We *shall reign for ever and ever* gives credence to the claim that we will be kings and priests for all
eternity; reigning over the renovated universe in constant, unending, joyful activity. We shall live by the ever-flowing river
of life, being constantly infused with new power and energy.

• **Will we marry?**
*For in the resurrection they neither marry, nor are giv-
en in marriage, but are as the angels of God in heaven*
(Matt. 22:30).

Will we marry? It all depends on who you ask. The above
verse seems to definitively indicate an unequivocal *no*, for did
not Christ Himself say that angels don't marry in heaven,
therefore, neither will we?

But on the other hand, many believe otherwise. They say
that Jesus indicated that the angels cannot *get* married in
heaven, but if they are already married in heaven, then they
will stay married. This, they say, also applies to us in the resurrection. Those who are not married in this life, or those
whose earthly marriage was something less than that which
Christ approved of, they cannot *get* married in heaven, but
those who experienced a true Christian marriage will enter
heaven married to the same earthly partner. Personally, I prefer to stay married in glory to my beloved wife.

CHAPTER THOUGHTS

How can you know completely of how the resurrection body
is to come from the natural body, or of the new heaven and the
new earth from the present heaven and present earth? What
words are there in your vocabulary that you may use to describe a spiritual body teeming with spiritual sDNA, or your
future existence in heaven, to be in the presence of the Lamb
of God, your Savior, your redeemer, and friend? What plati-

tudes are at your disposal to describe His omnipotence, His loving-kindness, His eternality?

We have done nothing for Him; He has done everything for us in the past and He will continue to do everything for us in the future. He will renovate the universe on our behalf; fill it with a new heaven and a new earth; construct the beautiful city of New Jerusalem in which to live, prepared as a bride adorned for her husband. There are no existing words that can adequately describe all that there is to be in heaven and in the New Jerusalem. It would be like describing how a giant oak tree can spring from a small acorn planted in the ground, a tree that you have never seen before, swaying majestically in the breeze. But it is a city, a glorious, majestic, and awe-inspiring city; the future home of the redeemed.

Cain built the first city, *Enoch* (Gen. 4: 17), without consultation with God, an enormous structure for that era but a complete failure. Sodom, Gomorrah, the beautiful cities of Babylon, and Rome, were all magnificent cities, but one thing they lacked; they built their cities without the blessings of God. Even today, our cities are great sores to our eyes and becoming more so as time advances. But the city of New Jerusalem shines in comparison with human endeavor, by its sheer size of 1500 miles on three of its dimensions, supported by jeweled foundations, solid pearl gates, and gold lined streets as pure and as clear as crystal. We shall never have a full understanding of what God has prepared for us until we are there, but one thing is sure, we shall all walk on the streets of that great city, and gaze upon the splendor and the beauty of our new home, for Jesus said: *In my Father's house are many mansions: . . . I go to prepare a place for you* (John 14:2).

Dear Christian friend, look for your name written in the Lamb's Book of Life! As angels escort you, as you arrive at your destination, your new home, New Jerusalem, enter through its pearled gates; always open; pause for a moment by the banks of the River of Life; listen to the music of the waters; He invites—whoever is athirst, come, quench your thirst, drink freely of the fountain of cool flowing waters; crystal pure as diamonds, flowing from the throne of God; gaze upon the Tree of Life on both sides; ponder all that befell you as you journeyed through life; its joys, its agonies, its pleasures and its heartaches; stroll along the city streets of pure gold, like unto clear glass; as the toes of your feet embrace the brilliance of its radiance, envision the wealth lavished upon you while the Savior suffered; giving His all for you at Calvary; a Man of sorrows, aquatinted with grief.

We go to a city as magnificent as pure gold; gaze upon its majestic walls, overflowing with jasper; its foundations garnished with all manner of precious stones, from jasper stone to amethyst; the Lamb is the Light of it; no need of the sun, neither of moon, to shine in it; behold the Sun of Righteousness; no more death, neither sorrow, nor pain; no more burdens of labor to bear, but peace and blessedness.

The message rings loud and clear—prepare thyself!—as the Bride, the Lamb's wife, adorned for her husband; get thee ready for the Marriage Feast of the Lamb; come ye blessed of My Father! He shall be our God and we shall be His sons; for the Lord God Almighty and the Lamb are the temple of it; boldly enter into the Holy of Holies; forever gaze upon His glory and majesty; for ye are a kingdom of priests, eternal priests in the heavens; officiating for the Great High Priest of the universe—Jesus the Christ—*Yeshua Hamashiach.*

These aspects of things to come are not just some sort of wishful thinking, or of delusional thoughts of fantasy, for they are all contained in the Book authored by the Holy Spirit. Dear Christian friend, make no mistake! Know this: There is an imminent Rapture coming, sooner than later; a Second Advent just over the horizon; a glorious, blissful 1,000 years of Millennial life abounding in love, joy, and peace for all those who claim the name of Jesus; a new heaven and a new earth teeming with life, stretching far out into the endless realms of eternity. Can you not now all shout—*Hallelujah!*

Dear friend, I wonder: *Do you REALLY know Him?*

24

HELL:
LOCATION! LOCATION! LOCATION!

---

*"Those who play with the devil's toys will be brought by
degrees to wield his sword"*
R. Buckminster Fuller

D on't believe in God? You are not alone! Yet, we would
be amiss at this point, nearing the end our discussion,
if we were to simply pass by the obvious fact that all
non-Christians, all those who have not bowed the knee to accept the Lord Jesus Christ as their personal Savior, to at least
bring to your attention, one more time, to convince you of your
complete ignorance of the most important fact in life. And that
fact is this: As the old saying goes, and to put it as bluntly as I
can—*"You are all going to Hell in a handbasket!"*

Now don't get angry with me. It is not I who is saying this.
The Bible minces no words when it speaks of divine judgment
for those who do not heed the warnings so plainly written. It
warns of impending disasters for neglecting the finished work
of Christ on the cross at Calvary—the death of the divine Son
of God for the sins of mankind—an act of infinite cost to an
infinite God; whose sole desire is that *all men to be saved, and
to come unto the knowledge of the truth* (1 Timothy 2:4).

Christians know exactly what I am trying to say to you.
Long ago, hell so scared me to look into the Scriptures to see
what this stuff was all about. I did not know anything about
predestination and election at that time; all I knew was that I
was afraid to go there and it was about time I did something
about it. What I did find is that, regardless of what others may

say about it, it was so easy to become Christian that it was really hard for me to believe it. All I really had to say was: "Thank you Lord for dying for me." That settled it for God and that's good enough for me. Reading further, I found out that when I die, I will live in the sky—literally! Don't believe it? Listen to what the apostle John has to say about it.

And I John saw the holy city, new Jerusalem, coming down from God out of heaven, prepared as a bride adorned for her husband (Rev. 21:2).

There are two things to notice here. The first is that all Christians will finally live in *the holy city, new Jerusalem, coming down from God out of heaven* and will remain somewhere out in space (I think), a beautiful city whose foundations and walls are of precious stones, *and the street of the city was pure gold* (v. 21). Secondly, it is a huge city whose dimensions are given as *twelve thousand furlongs* (v. 16), that is 1500 miles in length, breath, and height, enough room to accommodate all Christians throughout eternity. And the best thing of all is that the thrice holy God will live there: a*nd he will dwell with them, and they shall be his people, and God himself shall be with them, and be their God* (v. 3). It is safe to say at this point that life in the holy city will be awesome!

But for the wicked, there are no such promises in all of Scripture; no love of God; no rewards in heaven; no New Jerusalem to live in; no streets of gold to walk on; a new body but not in His image; no face-to-face contact with Him; no eternal life but an eternal death; not in the Lamb's book of life but in *the other* books; in hell and then in the lake of fire.

But the fearful, and unbelieving, and the abominable, and murderers, and whoremongers, and sorcerers, and idolaters, and all liars, shall have their part in the lake which burneth with fire and brimstone: which is the second death (Rev. 21:8).

Sorry, but you will not hear the call of the trumpet sound; no sun in your new abode, all blackness forever; lots more tears and sorrows; no temple of the living God; no river of water clear as crystal running through the streets; no Tree of Life; will never see Him; never talk with Him. What a disaster!

Enough has been said and written about the dismal existence of the unbelieving; those who have rejected His gracious
offer of salvation. Volumes have been written about the life
they are to experience after death, but what of their habitat?
What has been written about where they are to live? Where
are they going to spend eternity? In what place? In what land?
In what country? Don't you even want to know all that?

Realtors really understand what the word *location* means
in the sale of a property. If a mansion was advertised as the
most luxurious home in the world but was located in the middle of a desert as dry as a bone, or submerged up to your eye-
balls in a swamp full of alligators, would you still seek after it
for your final abode? Would you continue to strike a deal in
hopes of landing it? I truly believe that millions of you would,
and, sad to say, millions have already done so.

But it is not until the deal is cast in concrete and the day
for moving in is at hand, that the realization finally hits
home—"What have I done? Let me out of here!" But to your
chagrin you cannot back out of the contract. So you begin to
settle down and try to acclimate to the dire situation at hand.
But remember this—this is not a 30-year fixed mortgage contract that you're about to embark upon—it's forever!

BLACKNESS OF DARKNESS FOREVER

In order to accurately describe *that* place, we digress somewhat to lay down some background text and material to confirm our conclusions as to the nature and final destination of
the unbelieving; a home where even angels fear to tread.

It has been the topic for theologians ever since the days of
creation as to the nature and to the physical location of hell. Is
hell an actual location consisting of our three dimensional
space and time, or is it totally spiritual and void of all human
comprehension? Are there human souls there in eternal torment, or is it only a figment of some writer's imagination? Is it
a Biblical theology, a pagan philosophy, or an unintended consequence? And on and on it goes.

It has been conclusively identified by the Church down
through the ages, that hell is a real entity that is supported by
numerous Biblical passages, and, therefore, as occupying a
definite position in creation. But theories also abound as to

where the actual location of hell is. Is it located in the earth's center in the midst of the white-hot molten lava, as some Biblical passages seem to indicate? Is it located somewhere in the heavens surrounding the earth? Or might it be located in a Black Hole in some outer region of the universe?—all of which are conveniently tucked away somewhere out of sight.

The Bible does not specifically inform us as to the actual location since it is not necessary that we know of this fact, only that we are warned to abhor it altogether. We may, however, gain some insight as to the location of hell when we consider and apply sound Biblical exegesis with verifiable scientific facts and knowledge.

The nature of hell as described in the Bible is bleak and dreary. The one person who speaks mostly about hell in the Scriptures is our Lord Himself and, since He speaks with divine authority, we are to believe His revelation. His characterization of hell is as a temporary place of the departed souls of men who have rejected His gracious offer of salvation. It is the place of unsaved humanity between death and the Great White Throne judgment seat of Christ. The lost are said to *be cast out into outer darkness* (Matthew 8:12), where they are conscious, possess full use of their faculties and memory, and are in constant torment. There is no escape from it. This condition exists until the final judgment for then both Satan and those in hell will be *cast into the lake of fire* (Revelation 20:14). Jude describes the inhabitants of hell as:

Raging waves of the sea, foaming out their own shame; wandering stars, to whom is reserved the blackness of darkness forever (Jude 13).

Jude's description of hell (Heb. *Sheol*) is ominous and foreboding. He uses the metaphor of *blackness of darkness forever* to convey a hopelessness and despair of both heart and soul. Throughout the Bible, blackness and darkness are both synonymous with sin that characterizes every aspect of life as it exists there. The inhabitants in hell are willfully sinful and remain in that condition forever. However, I would propose that Jude's understanding of the phrase, "blackness of darkness forever," conveys a meaning far more important to this discussion than as to its common association with sin, for Job, centuries before, asked the question:

*Where is the way where light dwelleth? and as for
darkness, where is the place thereof, That thou
shouldest take it to the bound thereof, and that thou
shouldest know the paths to the house thereof?* (Job
38:19–20).

The Hebrew words for *dwelleth* and *place* both convey a
meaning of a land or a location of habitation. It seems as
though God is asking the question: Where is the *land of light*
and where is the *land of darkness*? And if you are to be a future inhabitant of one of these places, *you should know the
paths* to both lands since they are *bounded* and therefore imply definite *boundaries*. Certainly, there is more to this passage than just describing the light and darkness of morning
and evening that occurs when the earth rotates every twenty-
four hours on its axis. If God asks a question of us, we should
be diligent to search for an answer, for it cannot be beyond our
ability to do so. Job understood the question by confirming the
meaning of the Hebrew text, and gives us further insight as to
the nature of hell as the place of darkness; and he does so
without any mention or indication of its meaning as sin:

*Before I go whence I shall not return, even to the land of
darkness and the shadow of death; A land of darkness,
as darkness itself; and of the shadow of death, without
any order, and where the light is as darkness* (Job 10:
21–22).

THE CMBR

Turning to the scientific evidence as it relates to the subject of hell, it is an established scientific fact and a measurable
quantity, that the created universe contains what is known as
the CMBR, the *Cosmic Microwave Background Radiation*.
Admittedly, the discussion that follows is outside the area of
expertise of the writer, for the text matter lies in the scientific
area of Astrophysics, but finds justification for continuing this
line of reasoning for God has promised His assistance: *Come
now, and let us reason together, saith the Lord* (Isaiah 1:18).

The majority of evangelical Christians believe that God
created the universe "in place" and fully functional, in six literal days by an act *ex nihilo* (out of nothing), and is identified

as the "in-transit" cosmogony. *Cosmogony* is here defined as the study of creation during the six literal days, whereas *cosmology* is the study of the universe after the six days of creation, starting on the seventh day. Some Christian and secular scientists have advanced other systems of cosmogony that differ, some in a minor, and some in a radical sense. All cosmogonies do agree and confirm the existence of the CMBR, although some may arrive at different conclusions as to its origin. In order to reasonably explain the origin and nature of the CMBR, our discussion will focus on creation as starting from a White Hole cosmogony that preserves intact the six literal days of creation; the final result and effect of the CMBR being embraced by the majority of the differing cosmogonies.

The theory in a white hole cosmogony, first proposed by Humphreys,[1] which differs significantly from the evolutionary theories associated with the *black hole* origins of the universe, postulates that the earth is at or near the center of the universe; that this universe is bounded and has edges, similar to a flexible rubber ball. This is opposite of a black hole which postulates that the universe is unbounded and has no edges; that the earth is nowhere near the center of the universe because, as that theory asserts, there is no center of the universe.

Assuming a simplistic view at the instant of creation, at time $t = 0^+$, the universe consisted solely of pure energy and extremely hot, *without form and void* (Genesis 1:2, Heb. *tohu wabohu*). As the newly created energy expanded outward, and as one traveled with the expansion in a sphere of increasing radius R, the temperature *decrease* at any point along the radius was directly proportional to the distance *from* the center of the earth where the temperature was the hottest. As the universe *stretched out* with increasing radius R, the radiated energy stretched out from short wavelengths to longer wavelengths in relation to the stretching-out of the expansion.

It is he that sitteth upon the circle of the earth, and the inhabitants thereof are as grasshoppers; that stretcheth out the heavens as a curtain, and spreadeth them out as a tent to dwell in (Isa. 40:22).

As the energy cooled, stellar objects began to form in relation to Albert Einstein's Special Theory of Relativity, $E = mc^2$, where E is the energy of the universe that caused stars of mass *m* to form, a process similar to the energy/mass relationship in a nuclear reaction. All the created energy, E, however, did not materialize into stars. The final temperature of the leftover energy radiation, measured at the final radius R of the universe, at the edge of the sphere itself, and after everything was created that was to be created had finally cooled and settled down, is obtainable today at approximately 2.73 degrees as measured on the Kelvin temperature scale:

> Normal physical processes cause cooling to proceed as rapidly as the expansion. The stretching of space causes thermal ("heat wave") electromagnetic radiation in the expanse to drop from its initial very high temperature to much lower values in direct proportion to the increase in size of the expanse.[2]

This indicates that all throughout space in the universe there is an energy radiation in thermal equilibrium of temperature 2.73^0 K, and has come to be known as the CMBR, the *Cosmic Microwave Background Radiation*. This radiation permeates *all space* and is seen as approaching zero degrees Kelvin, absolute zero, wherein all activity in the universe stops. The universe is extremely cold—minus 460^0 Fahrenheit—except of course in the near vicinity of stars, such as our Sun.

ASTROPHYSICS 101

Radiation in the universe comes in all different flavors. The electromagnetic radiation spectrum, its frequencies and its wavelengths, lies between the values of zero and greater than 10^{23} Hertz (cycles per second), corresponding to wavelengths of infinity at direct current (DC) to less than 10^{-15} meters at gamma rays. Humans can visually see objects at wavelengths in a very narrow band at the midpoint range of 590 nanometers (10^{-9} meters), corresponding to color *yellow*. Ultraviolet rays are positioned directly above, and infrared rays are positioned directly below, the visible region. The total energy radiation spectrum, i.e., all the energy contained in the universe radiating at various wavelengths, is actually "light," free radiating photons, regardless of whether or not the human eye can visually see it. Our eyes are designed to filter out unwanted wavelengths, including ultraviolet and infrared light wavelengths, but pass visible light wavelengths that are impressed upon receptors in the retina and then processed in the brain.

A radio telescope, such as the Hubble Telescope that can electronically "see" radiation, in its Deep Field Probe image of portions of the universe, clearly shows regions glowing in infrared light. The CMBR, the leftover radiation from the creation of the universe, lies within the far-infrared portion of the microwave frequency range of the electromagnetic spectrum. The universe is full of infrared light emanating from this fundamental radiation and is the primary source of light in the universe, with a miniscule amount of visible light emanating from starlight.

Figure 24.1[3] is a microwave image of the *entire sky*, a 360-degree map using Galactic coordinates. This all-sky map is based on the first two years of data from NASA's Cosmic Background Explorer (COBE) satellite. After computer processing to remove contributions from Galactic emissions, nearby objects and the effects of the earth's motion, the map clearly shows the CMBR radiation distribution in our universe.

Now take away all the light from our Milky Way galaxy; take away all the light from remote stars and galaxies; take away all the glow of interstellar dust; what remains is a uniform, cosmic, infrared, background radiation—the CMBR!

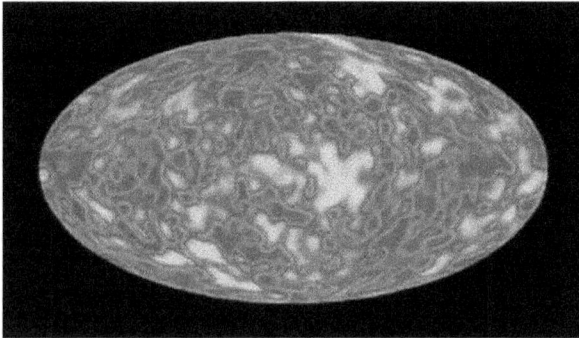

Figure 24.1: The CMBR

The Milky Way Galaxy, in which we all live, lies horizontally along the major axis and vertically along the minor axis, and is located at the mid-point of intersection. The computer generated colors represent radiation temperature variations, the dark regions (color red) indicating portions that are a hundredth of a percent warmer, and the lighter (color blue) indicating regions that are a hundredth cooler, than the average temperature of 2.73^0 K above absolute zero. The CMBR, *the Light of the world,* permeates all space with photon emissions! (Refer to Index No. 24.3 to view the CMBR in internet color).

> No matter where astronomers point their telescopes, they see a distant sheet of light surrounding us. Beyond that enormous shell of radiation, astronomers can see nothing. We are caged in by this surface; the cosmic microwave background (CMB).[4]

WHERE IS IT?

The Hubble Telescope can see stars in distant galaxies in the heavens in the form of starlight impinging upon the lens. Between the stars, and especially between galaxies, there are vast distances that are measurable in terms of hundreds of millions of light-years; the distance light travels in one year at 186,000 miles per second. The faint infrared light seen between the stars and galaxies *does not emanate* from the stars or galaxies themselves, but emanates from the CMBR. This phenomenon is akin to turning out all the lights in a bedroom only to find the walls, floor, and ceiling aglow with an eerie luminescence!

Beyond the Hubble Deep Field image astronomers can see *nothing!* There is no light emanating from beyond this image point. There is only *blackness!* There are no stars! There are no galaxies! There is no CMBR beyond this point! It seems that we have *seen the end of the universe of radius R!*

We may now draw the following conclusions:

- There is no *blackness of darkness within* the universe of radius R!—only light and light forever—the CMBR.
- There is no light *outside* of the universe of radius R since the CMBR ends at, and is contained within, the spherical surface of the universe.
- *Outside* of the spherical surface of the universe of radius R there is only *blackness of darkness forever.*

Science is the daughter of Theology and both are mutually reconcilable. Theology without science is wanting; science without theology is blind. Sound Biblical exegesis of the doctrine of hell, aided by modern scientific knowledge, combine to confirm the only location of hell that is tenable and which contain many infallible proofs. Hell does not lay within the center of the earth where the molten lava is the hottest and the brightest; does not have universal positioning coordinates in the highest of the heavens; nor does it reside within a light-filled black hole in some remote region in space.

Hell is located *outside of the sphere of the created universe of radius R!* The universe of radius R contains minute amounts of visible light and enormous amounts of infrared light in the form of the CMBR that permeates all of created space. There is no place in all of creation where there is no light! The Children of Light, those who name the name of Jesus as Lord and Savior, can only exist *within* the spherical surface of the universe of radius R, in an area of *light forever.* They cannot exist in an area of *blackness of darkness forever.*

The Children of Darkness, those who name the name of Satan as prince and ruler of this world, can only exist *without* the spherical surface of the universe of radius R, in an area of *blackness of darkness forever.* They cannot exist in an area of light forever. Quoting once again from Job, whose knowledge thousands of years ago absolutely knew of hell's location, speaking of lost souls at death and of their habitation:

He shall be driven from light into darkness, and chased out of the world (Job 18:18).

The Children of Darkness shall be chased out from living in the land of light, within the sphere of the universe, to living in a land of darkness, outside the sphere of the universe at the moment of death. They have no hope of returning to the land of light. The following Scripture now rings loud and clear:

Then spake Jesus . . . saying, I am the light of the world: he that followeth me shall not walk in darkness, but shall have the light of life. (John 8:12).

THAT GREAT GULF FIXED

The *great gulf fixed*, that great gulf that separates hell from paradise in Hades, where the Rich Man in hell is holding conversation with Abraham who is in paradise with Lazarus, a poor beggar during his earthly life:

> *And it came to pass, that the beggar died, and was carried by the angels into Abraham's bosom: the rich man also died, and was buried; And in hell he lift up his eyes, being in torments, and seeth Abraham afar off, and Lazarus in his bosom. . . . And beside all this, between us and you there is a great gulf fixed: so that they which would pass from hence to you cannot; neither can they pass to us, that would come from thence* (Luke 16: 22–23, 26),

that *great gulf fixed* is the outer sphere of the universe located at radius R! Children of Darkness can never cross over the great spherical gulf of the universe into the realm of the Children of Light. Sin has permeated all of God's creation and He will, in the final analysis, purge his creation with a New Heaven and a New Earth wherein sin and its remnants shall be no more; where *light forever* can never co-mingle with *blackness of darkness forever*. Their realm exists far out into the infinite blackness of darkness of nothingness, forever wandering and lost without hope; eternally separated from a Righteous and Holy God.

> *Yea, the light of the wicked shall be put out, and the spark of his fire shall not shine. The light shall be dark in his tabernacle, and his candle shall be put out with him* (Job 18:5–6).

The rich man further laments:

> *Father Abraham, have mercy on me, and send Lazarus, that he may dip the tip of his finger in water, and cool my tongue; for I am tormented in this flame. . . . Son, remember that thou in thy lifetime receivedst thy good things, and likewise Lazarus evil things: but now he is comforted, and thou art tormented* (Luke 16:24–25).

Through all this, I am reminded of the nineteenth century American poet, John Godfrey Saxe, who based the following poem on a fable told in India many years ago, entitled: *The Blind Men and the Elephant,* and in my estimation, bears a stark relationship to those who are of the Rich Men of today, those who have no ears to hear or eyes to see:

It was six men of Indostan
To learning much inclined,
Who went to see the Elephant
(Though all of them were blind),
That each by observation
Might satisfy his mind.

The *First* approached the Elephant,
And happening to fall
Against his broad and sturdy side,
At once began to bawl:
"God bless me! but the Elephant
Is very like a wall!"

The *Second*, feeling of the tusk,
Cried: "Ho!—what have we here
So very round and smooth and sharp?
To me 't is mighty clear
This wonder of an Elephant
Is very like a spear!"

The *Third* approached the animal,
And happening to take
The squirming trunk within his hands,
Thus boldly up and spake:
"I see," quoth he "the Elephant
Is very like a snake!"

The *Fourth* reached out his eager hand,
And felt about the knee.
"What most this wondrous beast is like
Is mighty plain," quoth he;
"'T is clear enough the Elephant
Is very like a tree!"

The *Fifth*, who chanced to touch the ear,
Said: "E'en the blindest man

> Can tell what this resembles most;
> Deny the fact who can,
> This marvel of an Elephant
> Is very like a fan!"
>
> The S*ixth* no sooner had begun
> About the beast to grope,
> Then, seizing on the swinging tail
> That fell within his scope,
> "I see," quote he, "the Elephant
> Is very like a rope!"
>
> And so these men of Indostan
> Disputed loud and long,
> Each in his own opinion
> Exceeding stiff and strong,
> Though each was partly in the right,
> And all were in the wrong!

The moral of this poem corresponds exactly to the behavior and attitude expressed by all modern-day Rich Men. They have eyes to see and ears to hear but retain little understanding of the things of God. *The Blind Men and the Elephant* serves as a warning, a lighted beacon in a raging sea of sin; of how our sensory perceptions can lead to disastrous misinterpretations that, in the final analysis, can lead us along the same path as that of the Rich Man, whose fate is eternally sealed:

> So, oft in theologic wars,
> The disputants, I ween,
> Rail on in utter ignorance
> Of what each other mean,
> *And prate about an Elephant*
> *Not one of them has seen!*

It was Charles Wesley who penned the following stanza in *"O For A Thousand Tongues To Sing."* Hear ye Him!

> *Hear him, ye deaf; his praise, ye dumb,*
> *your loosened tongues employ;*
> *ye blind, behold your savior come,*
> *and leap, ye lame, for joy.*

Dear friend, do not hesitate! Take heed! The time is now at hand. Be a Lazarus! Consider the Savior! Flee to the Cross and beg His forgiveness *ere what is written here falls heavily upon you*; to extinguish the light in your heart; to put out the flame in your soul.

CHAPTER THOUGHTS

Allow me at this juncture in our discussion to express my personal views with respect to the nature of the CMBR as I see it to be. Why was it designed to be there in the first place? Did God create too much energy back then? What function was it to perform? Who are those who might benefit from it?

The Bible speaks of three heavens. The 1st of the heavens encompasses the earth and the surrounding atmosphere where we all live and breathe; the 2nd heaven is outer space where stars, planets, and galaxies move about and have their being, both of which comprise the physical realm; and the 3rd heaven is where God resides, in the spirit realm. I confess that I don't know much about what's happening in the 3rd heaven.

Going back in time, however, what I do know is that prior to the creation of everything, there was only God. He did not live in a physical world at that time as there was no physical world at that time, only the 3rd heaven all over the place. He then opened a small hole in the sea of the 3rd expanse and dropped in the physical 1st and 2nd. He did not use the same atmosphere in the physical as there was in the spiritual. He then created the angels to be with Him in the 3rd. This is what we have today and this is where we all live.

But now, everything here seems to function differently from what's in the 3rd. When God speaks to those of us who are in the 1st heaven, a message from God to us on earth, something must exist at the boundary of the 3rd and the 2nd, where we go from the spirit realm into the physical realm. You would think that at this point the message contained in His thoughts would enter into a vacuum, a space entirely void of all matter, i.e., nothing there: no atoms; no stars, no galaxies, no light; no nothing. Well, that's not entirely true! There is something out there and God now works in and through what He has created.

If you studied physics and electrical theory in college, then you would know of such terms as resistance, capacitance, and inductance, to name but a few. Combining these three terms into one, we refer to them as *impedance*, a strange quantity that offers obstruction to all things, particularly to photon motion. That's the way God designed the physical universe; that's the way it is; and that's the way it will always be.

Engineers have grappled with this phenomenon for centuries. There is impedance out there in outer space that must be accounted for. There is no pure vacuum! Because of this, you can't simply stand on the boundary of the 3rd and the 2nd and yell at someone standing on the 1st and expect him to hear you. Impedance would see to it that the sound of your words go nowhere. Your message must *ride on something* in order to penetrate the impedance to reach everyone on the 1st.

Scientists have solved this by using inventions of radio and TV transmission theory where the message *modulates the frequency* on which it rides, to easily traverse the *"Characteristic Impedance of Free Space"*, Z_o, of 377 ohms. Now the message can travel thousands of miles suitable for reception on earth.

Enter now the world of the CMBR, that mysterious universe-wide spectrum of frequencies in the far-infrared region, designed solely for His purposes; existing in the 2nd and the 1st; always there, omnipresent, never missing a beat, ready to be modulated with the messages from God ordained for His people; light; free-radiating photons; 2.73 degrees Kelvin blackbody; 160.2 GHz center frequency, 1.873 mm wavelength; acting as a spiritual Post Office, delivering God's dThoughts (d for Divine) to each person, individually; a *universe wide web* (uww) delivery system, similar to the *world wide web* (www) system of today; entering the targeted body through the frontal cortex of the brain, producing dNeuropeptides that seek-out dReceptors of similar specificity; entering the cell to produce all that God intended man to be. *What an awesome God we have!*

Now then, hear this, when the Postman cometh for you, lift up your head on high, *for your redemption draweth nigh* (Luke 21:28). His message is the *Light of life,* information that modulates and rides upon the CMBR; itself photons of *light;* without which no message may be sent; no light to light-up your heart. You can't hide from it; you can't refuse to receive it; it will find you out by whatever means available:

The heavens declare the glory of God; and the firmament sheweth his handywork. Day unto day uttereth speech, and night unto night sheweth knowledge. There

*is no speech nor language, where their voice is not
heard. Their line is gone out through all the earth, and
their words to the end of the world. In them hath he set
a tabernacle for the sun, Which is as a bridegroom coming out of his chamber, and rejoiceth as a strong man to
run a race. His going forth is from the end of the heav-
en, and his circuit unto the ends of it: and there is nothing hid from the heat thereof* (Psa. 19:1–6).

This is what we know—that there is *speech* in all the heavens God has created, yearning to be heard by us. There is language out there; there is information out there; there is need
for understanding out there; from one end of the universe to
the other; waiting to be read; and more specifically, there is
life out there; and that only through the ministry of our Lord
Jesus Christ. *Hear ye Him!*

The front cover of this book depicts a ladder or staircase
coming out of heaven extending to earth, where the apostle in
John 1:50–51 proclaims: *thou shalt see greater things than these....Hereafter ye shall see heaven open, and the angels of God
ascending and descending upon the Son of man";* a representation of God communicating by means of angelic messengers,
and that via a ladder; a unique, pictorial passageway, speaking His thoughts *to* man and receiving prayers *from* man. This
is verified in Genesis 28 where God is definitely in communication with Jacob: *And he dreamed, and behold a ladder set up
on the earth, and the top of it reached to heaven: and behold the
angels of God ascending and descending on it. And, behold, the
Lord stood above it and said, I am the Lord God of Abraham,
thy father, and the God of Isaac . . .* (vvs. 12–13). God is definitely speaking to Jacob!

God's varied methods of communicating with man are also
pictorially revealed in Psalms 18:9–11 (then closely repeated
in 2 Samuel 22:10–12), wherein we may claim that God
warped the Time/Space domain to suit His purposes:

*He bowed the heavens also. And came down and dark-
ness was under his feet. And he rode upon a cherub,
and did fly; yea, he did fly upon the wings of the wind.
He made darkness his secret place; his pavilion round
about him was dark waters and thick clouds of the
skies* (vv. 9–11).

God is represented as descending down from heaven to earth at the cry of one of his children in distress. In our view, these verses clearly indicate that God actually came down *from* heaven *to* earth! How did He accomplish this? He simply *bowed* the heavens and came down!

Bowed (Heb. *natah*) means "to stretch out or thrust down into", or, "as parting a way to enter through" a portal, or tunnel, as through *dark waters and thick clouds.* This is not poetic license for skeptics to describe God's appearances but is a literal, *inter-dimensional travelling experience*, extending from the 3rd heaven, through both the 2nd and the 1st, to reach us on earth.

> *Bow thy heavens, O Lord, and come down; touch the mountains, and they shall smoke* (Psa. 144:5).

These passages apply to God, angels, and to various other entities (might it also relate to a Wormhole containing that portion of the CMBR modulated by the message from God, divinely raised to a degree of Warp Time/Space in order to circumvent the barrier of photon speed as equal to the speed of light?). Anything other than this, I do not know (You will have to ask an Astrophysicist!).

Do you think all this is too fanciful? Well then, think on this. Scientists[7] in Russia in the 1920's have tested and ratified that photons are transmitted by *every cell in the human body*, a cell-to-cell communication system (talking to each other!), unparalleled, carrying messages to other cells with the same specificity! It's like every cell in the human body is a miniature universe all their own, with receptor addresses to deliver to.

Now I am forever persuaded that God, in His majestic splendor, ordained what was right in His sight to do. Because God is Love, He ordained that you and I should enjoy eternity with Him; to walk with Him, to talk with Him; to gaze upon the glorious face of Jesus; to sit at His feet; to hear and to see how the worlds were first created; of how we were made in His image; to behold all the wonders of life; to explore the outer reaches of the universe; and to listen amazed at the sound of their speech:

And they that be wise shall shine as the brightness of the firmament; and they that turn many to righteousness as the stars for ever and ever (Daniel 12:3).

TIDBITS OF PART VII
- Resurrection:
  - o Material, not immaterial.
  - o A body transformed;
    - ♦ Metamorphosed.
- In heaven and on earth, simultaneously:
  - o Who?
    - ♦ Jesus, John, Paul, and all Christians.
  - o John → *Come up hither, and I will show thee.*
  - o Paul → *Such an one caught up to the third heaven.*
- The Rapture!
  - o When?
    - ♦ When the trumpet sounds.
  - o Who?
    - ♦ sDNA, regenerated and resurrected.
- Phase 5: *Glorification → Adoption → New Birth → Sons:*
  - o Glorification;
    - ♦ Put on Christ!
  - o Adoption;
    - ♦ Waiting for the redemption of our body.
  - o New Birth;
    - ♦ Bear the image of the heavenly.
  - o Sons;
    - ♦ Await the finality of the resurrection.
- Heaven:
  - o Where?
    - ♦ In the land of Light.
  - o New Jerusalem;
    - ♦ Our new home.
    - ♦ Great walls, foundations, and names.
  - o Night, sun, candles there?
    - ♦ No!
  - o Tears, death, sorrow, crying, or pain there?
    - ♦ No!
  - o The Tree of Life there?
    - ♦ Yes!
  - o Who will be there?

- ♦ You!
- ♦ The Bride of the Lamb of God.
  - o Married to your wife?
    - ♦ Depends on who you ask!
  - o Will Jesus be there?
    - ♦ YES!
    - ♦ His Signature is on the "face of the deep".
    - ♦ *The Light of the World!*
- The CMBR:
  - o What is that?
    - ♦ Leftover creation energy.
  - o Energy in what form?
    - ♦ Photons.
  - o What are photons?
    - ♦ Electromagnetic radiation.
  - o You mean basically light?
    - ♦ Yes.
  - o How fast is it?
    - ♦ 186,000 miles per second.
  - o Isn't that too slow?
    - ♦ Yes, but it's warped.
  - o What can we use it for?
    - ♦ As a signal transmission medium.
  - o What kind of signal?
    - ♦ Messages from God.
  - o To who?
    - ♦ To You and to Me!
- Hell:
  - o Blackness of darkness forever.
  - o Where is that?
    - ♦ Beyond the universe of radius "R".
  - o WHERE?
    - ♦ Beyond the Great Gulf fixed!
  - o Where is that?
    - ♦ Beyond the CMBR!
  - o Who will be there?
    - ♦ All those who refuse Christ's forgiveness!
  - o Will Jesus be there?
    - ♦ NO!

Parting Thoughts

Consider This Man JESUS!

Dear Christian friend, if you never committed passages of Scripture to memory, or have never highlighted a study chapter in your Bible, then I would suggest to you that you do so with the Epistle to the Hebrews. This epistle in my estimation is the Gospel expressed in a "nutshell". The author out-did himself when he penned the words, comparing Jesus to anything or to anyone in all of creation that we may think is better than He is.

The author (I think it to be the apostle Paul) stresses the obvious fact that God is speaking! God is continually speaking to us. The epistle opens with God, firstly, speaking to the O.T. saints *in time past,* but now He speaks to us *in these last days.*

Know this: A silent God is an unknown God. God speaking is God revealing Himself, and this is done through His Word. Jesus Christ is the language of the message to men; the message is in God's Word, and Jesus is the Word.

Allow me to list those things that stand out in the contents of God speaking in Chapter 1 of the Epistle to the Hebrews:

I. God is speaking *by his Son* (v. 1):
  • The Person who *spake* was *God* (the Father).
  • The Procedure was *at sundry times and in divers manners* (at different times and in different ways).
  • The Period was in the O.T., *in time past.*
  • The *Fathers* spoken to was the Jewish nation.
  • The *Prophets* spoken to were those who spoke for God.
II. God is speaking *in* His Son (vv. 2–4):
  • The Period of the Son is *in these last days* (Now!).
  • Jesus is the message!
  • He is the *Inheritor* of all things.
  • He is the *Creator* of all things.

- He is the *Upholder* of all things.
- He is the *Brightness* of God's glory.
- He is the *Express Image* of God's person.
- He by Himself *Purged Our Sins.*
- He sits *on the Right Hand* of God the Father.
    o He occupies a *Rest* position (nothing else need be done).
    o He occupies a *Royal* position (He is God).
    o He occupies a *Renowned* position (He died for our sins).
- He is *so much better than the angels.*

III. God is speaking *for* His Son (vv. 5–7):
- God spoke for His Son in His Resurrection: *Thou art My Son, this day have I begotten thee.*
- God spoke for His Son in His Relationship: *I will be to him a Father, and he shall be to me a Son.*
- God will speak for His Son in His Return: when Jesus returns *again* to Earth.
- God said; *Let all the angels worship Him.*

IV. God is speaking *to* His son (vv. 8–14):
- When God says: *Thy throne,* it indicates His Majesty.
- When God says: *O God,* it indicates His Deity.
- When God says: *Forever and ever,* it indicates His Eternality.
- When God says: A *scepter of righteousness,* it indicates His Authority.
- When God says: *But thou art the same,* it indicates His Immutability.
- When God says: *I make thine enemies thy footstool,* it indicates His Superiority.

This reminds me, at this point, of a sermon by the late Dr. S. M. Lockeridge, pastor of Calvary Baptist Church, San Diego, CA, given in 1976 while in Detroit, MI, entitled: *That's My King,* emphasizing the beauty, the righteousness, and the glory of our Lord. I can just imagine the enthusiasm, the joy, and the expectations of those sitting in the congregation; listening attentively for what is to follow, for it speaks of, and emphasizes, all that is listed above, and then on to numerous volumes thereafter. He writes:

My King was born King.
The Bible says He was a seven way King.
He's the King of the Jews - that's an Ethnic King.
He's the King of Israel - that's a National King.
He's the King of Righteousness. He's the King of the Ages.
He's the King of Heaven. He's the King of Glory.
He's the King of Kings. And He is the Lord of Lords.
Now that's my King!

Well, I wonder if you know Him. Do you know Him?
Don't try to mislead me. Do you know my King?
David said The Heavens declare the glory of God
   And the firmament showeth His handiwork.
No means of measure can define His limitless love.
No far seeing telescope can bring into visibility the coastline of
   The shore of His supplies.
No barriers can hinder Him from pouring out His blessing.
He's entirely sincere. He's eternally steadfast.
He's immortally graceful. He's imperially powerful.
He's impartially merciful.
That's my King!

He's God's Son. He's the sinner's Saviour.
He's the centerpiece of civilization.
He stands alone in Himself.
He's honest. He's unique. He's unparalleled.
He's unprecedented. He's supreme. He's pre-eminent.
He's the loftiest idea in literature.
He's the highest personality in philosophy.
He's the supreme problem in higher criticism.
He's the fundamental doctrine in historic theology.
He's the cardinal necessity of spiritual religion.
That's my King!

He's the miracle of the age.
He's the superlative of everything good that you call Him.
He's the only one able to supply all our needs simultaneously.
He supplies strength for the weak.
He's available for the tempted and the tried.
He sympathizes and He saves. He guards and He guides.
He heals the sick. He cleanses the lepers. He forgives sinners.

He discharges debtors. He delivers the captives.
He defends the feeble. He blesses the young.
He serves the unfortunate. He regards the aged.
He rewards the diligent. He beautifies the meek.
That's my king.

Do you know Him?
Well, my King is the King of knowledge.
He's the wellspring of wisdom. He's the doorway of deliver-
ance.
He's the pathway of peace. He's the roadway of righteousness.
He's the highway of holiness. He's the gateway of glory.
He's the master of the mighty. He's the captain of the conquer-
ors.
He's the head of the heroes. He's the leader of the legislators.
He's the overseer of the overcomers.
He's the governor of governors. He's the prince of princes.
He's the King of Kings. And He's the Lord of Lords.
That's my King.

His office is manifold. His promise is sure.
His life is matchless.
His goodness is limitless. His mercy is everlasting.
His love never changes. His word is enough.
His grace is sufficient. His reign is righteous. His yoke is easy.
His burden is light. I wish I could describe Him to you,
    But He's indescribable.
That's my King.

He's incomprehensible. He's invincible. He's irresistible.
I'm coming to tell you
    That the heaven of heavens cannot contain Him,
    Let alone a man explain Him.
You can't get Him out of your mind.
You can't get Him off of your hands.
You can't outlive Him and you can't live without Him.
The Pharisees couldn't stand Him
    But they found out they couldn't stop Him.
Pilate couldn't find any fault in Him.
The witnesses couldn't get their testimonies to agree
    And Herod couldn't kill Him.

Death couldn't handle Him and the grave couldn't hold Him.
That's my King!

He always has been and He always will be.
I'm talking about the fact that He had no predecessor
    And He'll have no successor.
There's nobody before Him and there'll be nobody after Him.
You can't impeach Him. And He's not going to resign.
That's my King!

Thine is the Kingdom. And the power. And the glory.
Well, all the power and glory belong to my King.
We're around here talking about black power
    And white power and green power.
In the end all that matters is God's power. Thine is the power.
    Yeah, and the glory.

Thine is the Kingdom. And the power. And the glory.
Forever.
    And ever. And ever. And ever.
How long is that?
Forever
    And ever. And ever. And ever.
And when you get through with all of the Evers.
    Then . . . Amen.

Please, allow me this one last time to tell you about this
Man, Jesus, of how He has had His way with me, personally,
and possibly with some of you also. Read carefully the following, from the works of G. D. Watson, a Methodist minister
(1845-1924): *Others May, You Cannot;*

"If God has called you to be really like Jesus, He will
draw you into a life of crucifixion and humility, and put
upon you such demands of obedience, that you will not
be able to measure yourself by other Christians; and in
many ways, He will seem to let other good people do
things which He will never let you do.

Other Christians and ministers, who seem very religious and useful, can push themselves, pull wires and
work schemes to carry out their Christian goals, but

these things you simply cannot do. Others may boast of their work or their writings or their success, but the Holy Spirit will not allow you to do any such thing, and if you ever try it, He will lead you into some deep mortification that will make you despise yourself and all your good works.

Others may be allowed to succeed in making money, but most likely God will keep you poor, because He wants you to have something far better than gold, namely, a helpless dependence on Him and the joy of seeing Him supply your needs day by day out of an unseen Treasury.

The Lord may let others be honored and keep you hidden and unappreciated because He wants to produce some choice, fragrant fruit for His coming glory, which can only be produced in the shade. He may let others do a work for Him and get the credit for it, but He will make you work on and on without others knowing how much you are doing; and then, to make your work still more precious, He may let others get the credit for the work which you have done, and thus make your reward ten times greater when Jesus comes.

The Holy Spirit will rebuke you for little words or deeds or even feelings, or for wasting your time, which other Christians never seem to be concerned about, but you must make up your mind that God is an infinite Sovereign and He has a right to do whatever He pleases with His own. He may not explain to you a thousand things which puzzle your reason in the way He deals with you, but if you will just submit yourself to Him in all things, He will wrap you up in a jealous love and bestow upon you many blessing which come only to those who are very near to His heart.

Settle it then, that He is to have the privilege of tying your tongue, or chaining your hand, or closing your eyes, in ways that He does not seem to use with others. Now, when you are so possessed with the living God that your secret heart becomes pleased and delighted

with this peculiar, personal, private, jealous guardian-
ship and management of the Holy Spirit over your life,
then you will have **entered the very vestibule of
heaven itself.**"

I am so sorry but at this point I am obliged to leave you to
your own thoughts. If you have someone in mind that has
equaled or surpassed what has been written above, or
throughout this entire volume, in all recorded history, please
do not hesitate to identify him to us that we may know and
applaud him for his achievements (but I seriously doubt that
you can!).

Ponder endlessly what has been written herein. Do not be
dismayed. Be strong. Be assured that I know each of you intuitively, and that each of you has some remaining doubts in your
mind as to the things that you have read, but may have never
heard before. Look to Jesus! Ponder the Scriptures! Hear Him
when He speaks! Follow Him when you are invited! It will
then be revealed to you of what it is for you to know and to do.

A few solemn words for our non-Christian friends: The
Lord has blinded you to spiritual realities because you have
rejected the Scriptures that speak of Him of what He has accomplished at the Cross. How different are Christians as we
look at the very same Scriptures, for in them I see the beauty
and majesty of Almighty God in all His Glory, manifested in
the person of Jesus. But when you look at the very same Scriptures, all you can see, as do the blind men, is anything in your
imagination other than that of Christ.

Is this your station in life? If so, which of the six blind men
are you as you look into the Scriptures but see the elephant?
Do you see a *wall,* a *spear,* a *snake,* a *tree,* a *fan,* a *rope?* Whatever you read, see, or hear of, concerning Christ you immediately label as wrong and contrary to the facts as you perceive
them. But the blind men attest to your ignorance and loudly
proclaim to all of you: *"And all were in the wrong."*

Flee to Christ and beg forgiveness that He may heal your
eyes that you may see, that you may at last gaze upon and
come to know the Christ that we know, not a god of sticks and
stones, of bricks and mortar, but a God of love, joy, and peace.
May the Holy Spirit grant to you the grace to do so.

Every Christian living on earth should be thrilled to sing the lyrics to that grand old song, sung by the legendary tenor, "Tennessee" Ernie Ford—*Come On Down*:

> *Come on down, Lord Jesus, and take us away,*
> *Come on down, Lord Jesus, could this be the day?*
> *Even so, come quickly, Lord Jesus we pray,*
> *Come on down, Lord Jesus, come soon.*

Raise the banners up on high! Sound loud the clarions!
I am *ransomed, healed, restored, forgiven!*
Praise God! Praise God!—I'm a Child of the KING!

I have many things to say to you about Jesus, but these must wait for another day. In the meantime, seal-up the words!

> *And there are also many other things which Jesus did, the which, if they should be written every one, I suppose that even the world itself could not contain the books that should be written. Amen* (John 21:25).

How else could I have loved you, O my GOD!

INDICES

Index of Scripture

Index of Endnotes

Chapter 2: Leaves, Worms, Butterflies & T. U. L. I. P. S. (Page 33).
1. Mayes, Virgil C., *Leaves, Worms, Butterflies, and T.U.L.I.P.S.,* (Splendora, Texas, 1979), p. xii.
2. Boettner, Loraine, *"The Reformed Doctrine of Predestination,"* (Phil., PA, Presbyterian and Reformed Pub. Co., 1932, 1980), p.433-436.

Chapter 3: Two Inviolable Laws (Page 41).
1. Asimov, Isaac, *In the Game of Energy and Thermodynamics You Can't Even Break Even,* Journal of Smithsonian Institute, June 1970, p.6.

Chapter 4: In the Beginning–DNA (Page 55).
1. Figure 4.1: *DNA Molecule,* obtained from website: http://www.proprofs.com/quiz-school/story.php?title=dna-mutation-mitosis
(Science fact, Author unknown).
2. Figure 4.2: *Genes Are Us,* obtained from website: http://www.nature.com/scitable/topicpage/chemical-structure-of-rna-348
(Science fact, Author unknown).
3. Figure 4.3: *From Gene To Protein,*
(Science fact, Author unknown)
4. Figure 4.4: *Chirality,* obtained from website: http://commons.wikimedia.org/wiki/File:Chirality_with_hands.jpg. (Author unknown)

Chapter 5: From DNA to People (Page 65).
1. Figure 5.1: *Transcription of mRNA,* obtained from website: http://gene-tics.wikispaces.com/17.+From+Gene+to+Protein
(Science fact, Author unknown).
2. Figure 5.2: *Translation of mRNA,* obtained from website: http://www.cancer.gov/cancertopics/understandingcancer/genetic variation/page22
(Science fact, Author unknown).
3. Figure 5.3: *The Cell,* obtained from website: http://commons.wikimedia.org/wiki/File:Cell_parts.png

Chapter 6: Biological Immortality (Page 73).

1. (Author unknown).
2. Figure 6.1: *Telomere Shortening,* obtained from website: www.stemcells.nih.gov/info/scireport/appendixc.asp (Author unknown).
3. Skloot, Rebecca, *Henrietta's Dance:* (John Hopkins Magazine, April 2 000), obtained from the website: http://www.jhu.edu/~jhumag/0400web/01.html

Chapter 7: Design! Design! Design! (Page 81).

1. Figure 7.1, *The Nervous System,* obtained from website http://ais-humanbiology.wikispaces.com/Nervous+system (Author unknown).
2. Denton, Michael, *Evolution: A Theory in Crisis* (London: Burnett Books, Ltd., 1985), p. 330.
3. Figure 7.2, *The Nerve Cell,* obtained from website: http://commons.wikimedia.org/wiki/File:Neuron.svg (Author unknown).
4. Glassey, Donald, *The Soul Swims in the CSF;* The Metaphysical Physiology of the Cerebrospinal Fluid, obtained from the website: http://www.healtouch.com/csft/soul.html
5. Figure 7.3, *Cell Communications,* obtained from website: www.comons.wikimedia.org/wiki/File:1Signal_Transduction_Path h-ways_Model.jpg (Author: Yaneporn).
6. Secko, David, The Science Creative Quarterly, *"Conversing at the Cellular Level,"* No. 6, 2011.
7. Pert, Candace, *"Molecules of Emotion: Why You Feel the Way You Feel"*, (N.Y., Scribner, 1987), p. 23.
8. Ibid, p. 164.
9. (Author unknown).

Chapter 8: Do You Know Him? (Page 95).

1. Luginbill, Robert. D., *Bible Basics: Essential Doctrines of the Bible,* Part I, Section II C3: The Trinity in the Old Testament, obtained from website: http://www.ichthys.com/1Theo.htm, Part I. Section II, The Persons of God: The Trinity.
2. Weismann, August, *The Germ Plasm: A Theory of Heredity,* Part 2, Ch. 6, translated by W.N.Parker, (Scribner's Sons, NY, *1893),* p. 183ff.
3. Figure 8.1: Hardy, Alister, *The Living Stream,* (Collins, London, 1965), p.76.
4. Pink, Arthur, *Exposition of Hebrews,* (Baker Book House, Grand Rapids, MI 1954) p. 341.

Chapter 9: Murderer! Who? *Me?* (Page 115).
1. Strong, James, *Strong's Concordance*, (CB Pub., Nashville, Tn) #5061 & #5221.
2. Spurgeon, C.H., *The Treasury of David*, Vol.1, (MacDonald Pub. Co., McLean, VA). p328.
3. Ibid.

Chapter 10: Biological Regeneration (131)
1. Webster's II New Riverside Univ. Dictionary, *Metamorphosis*," (Riverside Pub. Co., 1913) p. 745.
2. Figure 10.1: *Virus Attack*, obtained from website: http://commons.wikimedia.org/wiki/File:Virus_Replication.svg (Author unknown).

Chapter 11: The New Birth Zygote (Page 145).
1. Brooker, Robert J., *Genetics, Analysis and Principles,*" (Benjamin Cummings Pub. Co., 1998).
2. Nee, Watchman, *Regeneration Is a Matter of Life*, reproduced from the website: http://ww.regenerated.net/life/index.html
3. Hyles, Jack, *Salvation Is More Than Being Saved*, Chapter 2, Regeneration and Salvation, reproduced from the website; http://www.jesus-is-savior.com/Books,%20Tracts%20&%20Preaching/ Printed%20Books/Dr%20Jack%20Hyles/Salvation/salvation-index.htm
4. Cole, Steven J., *Born Again to Love,* obtained from the Flagstaff Christian Fellowship website: http://www.fcfonline.org/content/1/sermons/072692m.pdf

Chapter 12: Love's Long Journey (Page 163).
1. Figure 12.1, *Nucleotide Mutations,* obtained from website: http://commons.wikimedia.org/wiki/File:DNA_UV_mutation.gif (Author unknown)
2. Brooker, Robert J., *Genetics, Analysis and Principles,* (Benjamin Cummings Pub. Co., 1998), p.6.
3. Figure 12.2, *Holographic Recording,* obtained from website: http://commons.wikimedia.org/wiki/File:Holography-record.png. (Author unknown).
4. Obtained from website: http://en.wikipedia.org/wiki/Holography. (Author unknown).

Chapter 13: Biological Armageddon (Page 177).
1. Bradford, John, *The Old Man & The New,* (William Serres, London, 1567), p. 297
2. Ibid, p. 298

3. Jones, John Cynddlan, *Studies in the Gospel According to St, John,* (Hamilton, Adams & Co., London, England), p. 198.

4. Figure 13.1, *White Blood Cell Lineage,* obtained from the website:
http://www.wikicell.org/index.php/Bone_Marrow_Progenitor_Cell.
(Author unknown).

Chapter 14: Love, Joy & Peace (Page 191).

1. Garner, Patrick, *Biotechnology and the Connection Between Our Physical and Spiritual Natures,* The Center for Bioethics and Human Dignity, obtained from the website:

2. Pert, Candace, *Molecules of Emotion: Why You Feel the Way You Feel,* (N.Y., Scribner, 1987).

3. Meier, Paul, *Brain Chemicals Linked To Physical and Emotional Health,* obtained from the website:
http://www.meierclinics.com/xm_client/client_documents/Ra
dioHandouts/Brain_Chemicals_Linked_to_Phy__Emot_Heal
th.pdf

Chapter 15: The New Birth Blastocyst (Page 203).

1. Pink, Arthur. *"Spiritual Growth"* obtained from the website:
http://www.godrules.net/library/pink/246pink0.htm

2. Clarke, Adam, *Clarke's Commentary-Psalms 85,"* obtained from the website:
http://www.godrules.net/library/clarke/clarkepsa85.htm

3. Pink, Arthur, *Spiritual Growth—Its Analogy,* obtained from the website:
http://www.pbministries.org/books/pink/Spiritual_Growth/growt
h_05.htm

4. Figure 15.1, *Blastocyst,* obtained from website:
http://www.en.wikipedia.org/wiki/File:Blastozyste.svg
(Author unknown).

5. Pink, Arthur, Ibid.

6. Wikipedia Chimera article, obtained from website:
http://www.wikipedia.org/wiki/Chimera_(genetics)

Chapter 16: Light's Shinning Glory (Page 215).

1. Spurgeon, Charles, *Psalms 139,* The Treasury Of David, Vol. III, (MacDonald Pub. Co., McLean, Va.) p. 260.

2. Biblia Hebraica Stuttgartensia, ed. R. Kittle, *Psalms 139:16,* p.1218.

3. Strong's Concordance, *Logos,* #3056.

4. Adam Clarke's Commentary, Vol.V, *Matthew-Acts,* (Methodist Book Concern, NY).

5. Pink, Arthur, *The Gospel of John,*
Http://www.pbministries.org/books/pink/John/john.htm

6. Strong's Concordance, *phos, phemi,* #5457, #5433.

Chapter 18: The New Birth Embryo (Page 239).

1. Pink, Arthur, *Spiritual Growth—Its Analogy,* reproduced from the Providence Baptist Ministries website: http://www.pbministries.org/books/pink/Spiritual_Growth/growth_05.htm

2. Johnson, S. Lewis, *In the Light–A Three-fold Assurance,* reproduced from the SLJ Institute website: http://www.sljinstitute.net/sermons/new%20testament/general/pages/123john8.html

3. Pink, Arthur W., *An Exposition of Hebrews,* (Baker Book House, Grand Rapids, MI, 1954), p. 378.

4. Clarke, Adam, Clarke's Commentary—Galatians 3, obtained from website: http://www.godrules.net/library/clarke/clarkegal3.htm

5. Clarke, Adam, *"Clarke's Commentary—Romans 13,"* obtained from website: http://www.godrules.net/library/clarke/clarkerom13.htm

Chapter 19: Life's Amazing Story (Page 253).

1. Wikipedia, *Butterfly,* http://www.wikipedia.org/wiki/Butterfly

2. Adam Clarke's Bible Commentary, *John 6,* http://www.godrules.net/library/clarke/clarkejoh6.htm

3. Spurgeon, C.H., *Light At Evening Time,* http://www.biblebb.com/files/spurgeon/3508.txt

4. Adam Clarke's Bible Commentary, *Ezekiel 36,* http://www.godrules.net/library/clarke/clarkeeze36.htm

Chapter 20: Faith, Meekness & Temperance (Page 265).

1. Pettinelli, Mark, *The Psycology of Emotions, Feelings, and Thoughts,* p. 28. obtained from website: <cnx.org/content/m14358/latest/>.

2. Pert, Candace, *Molecules of Emotion: Why You Feel the Way You Feel,* (N.Y., Scribner, 1987), p. 189.

3. Chopra, Deepak, *Quantum Healing,* (Bantam Books, NY, 1989), p. 95.

4. Chisholm, Thomas O. and Runyan, William M., *Great Is Thy Faithfulness,* 1923.

Chapter 21: The New Birth Fetus (Page 277).

1. Huckabee, Davis W., *Studies On Strong Doctrine—Sanctification,* obtained from the PBCMinistries website: http://www.pbministries.org/Theology/Davis%20Huckabee/To%20Studies%20In%20Strong%20Doctrine/strong_doctrine_11.htm

2. Spurgeon, Charles H., *Threefold Sanctification,* Metropolitan Tabernacle Pulpit, 1862 Vol. 8, Sermon 434, obtained from website: http://www.spurgeongems.org/vols7-9/chs434.pdf

3. Pink, Arthur, *Spiritual Growth*, obtained from website:
    <godrules.net/library/pink/246pink0.htm>.
4. Clarke, Adam, *Clarke's Commentary—1 John 2*, obtained from
    the website:
    http://godrules.net/library/clarke/clarke1joh2.htm
5. Scofield, C.I., The New Scofield Reference Bible, 1 Pet. 2:9, Note
    1, p.1334.
6. Figure 21.1, *Wilderness Tabernacle,* obtained from website:
    http://commons.wikimedia.org/wiki/File:Foster_Bible_Pictures_0
    071-1_The_Tabernacle_that_the_Israelites_Built.jpg
    (Author unknown).
7. Figure 21.2, *The Human Cell,* obtained from website:
    http://commons.wikimedia.org/wiki/File:Biological_cell.svg.
    (Author unknown)

Chapter 22: Biological Resurrection (Page 295).

1. Hirt, Herb, *The Lie,* Israel My Glory Magazine, Vol.65, No.4, Ju-
    ly/August 2007, The Friends of Israel Gospel Ministry.
2. Pyles, David, *The Spiritual Body,* reproduced from The Primi-
    tive Baptist website:
    http://www.pb.org/pbdocs/spiritual_body.html
3. Matthew Henry's Commentary, *Matthew To John,*
    (F.H. Revell Co., London), Vol. V, p. 886.

Chapter 23: The New Birth: The Sons of God (Page 307).

1. Wayne, Grudem, *Systematic Theology: An Introduction to Bibli-
    cal Doctrine,* (Zondervan Publishing and InterVarsity Press,
    1994), p. 828.
2. Chafer, Lewis Sperry, *Systematic Theology,* Vol II, p.155
3. Scofield Bible 1917, Ephesians 1:5, Note 4.

Chapter 24: Hell: Location, Location, Location (Page 325).

1. Humphreys, D. Russell, *Starlight and Time,* (Master Books,
    Green Forest, AR 1994).
2. Ibid, p. 77.
3. Figure 24.1, *The CMBR,* obtained from website:
    http://commons.wikimedia.org/wiki/File:COBE_cmb_fluctuations
    .gif.
    (NASA, Author unknown).
4. Seife, Charles, *Peering Backward to the Cosmos Fiery Birth,* Sci-
    ence (vol. 292, June 22, 2001), p. 2236.

(Note: Websites listed above are valid for as long as they are maintained active by the Webmaster).

Index of Subjects

www.ingramcontent.com/pod-product-compliance
Lightning Source LLC
Chambersburg PA
CBHW061559110426
42742CB00038B/1594